FOOD PACKAGING MATERIALS

Aspects of Analysis and Migration of Contaminants

FOOD PACKAGING MATERIALS

Aspects of Analysis and Migration of Contaminants

N. T. CROSBY

Head of Food Additives, Contaminants and Information, Laboratory of the Government Chemist, London

APPLIED SCIENCE PUBLISHERS LTD
LONDON

APPLIED SCIENCE PUBLISHERS LTD
RIPPLE ROAD, BARKING, ESSEX, ENGLAND

British Library Cataloguing in Publication Data

Crosby, N T
Food packaging materials.
1. Food contamination 2. Plastics in packaging 3. Plastics—Additives—Analysis 4. Food—Analysis
I. Title
363.1′9264 TX571.P63

ISBN 0-85334-926-6

WITH 39 TABLES AND 21 ILLUSTRATIONS

Printed in Great Britain by Galliard (Printers) Ltd, Great Yarmouth

ACKNOWLEDGEMENTS

I wish to thank the Government Chemist for permission to publish this manuscript. The views expressed are entirely personal and do not necessarily represent those of the Laboratory of the Government Chemist, the Department of Industry or Her Majesty's Government.

I thank my present and former colleagues for their help in the preparation of this manuscript.

I acknowledge permission to reproduce certain Figures and Tables as follows:

Chapter 3, Fig. 1—International Scientific Communications Inc.
Chapter 3, Figs. 3, 5, 6 and 7 and Chapter 7, Figs. 1, 2, 3 and 4—Pergamon Press Ltd.
Chapter 5, Table 2—The British Ceramic Research Association.
Chapter 6, Fig. 3 and Chapter 7, Fig. 6—Morgan-Grampian (Process Press) Ltd.
Chapter 8, Fig. 2—Springer–Verlag New York Inc.

PREFACE

Whilst glass, tinplate and paper have been widely used as food packaging materials for many years, plastics are a relatively recent innovation in this field. Major technological changes can introduce many benefits to mankind but at the same time they are liable to create new problems. Public concern over the use of chemicals and their release into the environment has increased markedly in the last ten years. This concern has been matched by a growth in legislative controls to protect the public in the home, at work and by general restrictions on pollution of the environment. No area is of greater importance than the nation's food supply. Hence the use of new chemicals, even though only in contact with food, as opposed to direct consumption as food additives, has to be examined with great care. Nevertheless, the development of the plastics industry has been a major technological achievement, and applications in the field of food packaging have contributed considerably to the conservation of scarce resources, so vital when the world's population is still increasing at a rapid rate. Furthermore, in protecting the food from contamination by outside chemical or microbiological agents, plastics assist in the development and maintenance of better food hygiene practices; the major area of concern at the present time.

Although plastics (i.e. the polymers) themselves are generally inert, monomers and chemicals used as processing aids may also be present in the finished product. Such chemicals may transfer from the plastic into the food during storage. Although not all such chemicals are harmful, there is a need to restrict the ingress of unnecessary non-nutritive substances into food. This has led to increasing activity from regulatory authorities, analytical chemists and toxicologists in an attempt to define and measure the migration problems and to quantify the risk arising from the use of plastics as food packaging materials. There is clearly no risk to the consumer even from harmful chemicals unless they are actually transferred from the plastic into the food. Hence, theoretical and experimental studies have been pursued to elucidate both the mechanism and the extent of the transfer process.

The science of food packaging is, therefore, truly multidisciplinary in so far as it encompasses parts of all the following activities: plastics and polymer technology, food science, physical chemistry, analytical chemistry, toxicology, national and international legislation—and probably many others too. Each of these subject areas has expanded rapidly in recent years so that an enormous volume of literature has accumulated. From this vast body of information, I have attempted to extract the relevant data likely to be of interest to both administrative and scientific staff in the food and plastics industries. Theoretical concepts and experimental evidence have been collated and assessed critically for the assistance of those working in industry, government or academic fields, who have undergone a basic training in science but have no specialist knowledge of food packaging. Parts of the book may also be of interest to students and those preparing for examinations such as the Mastership in Food Control. Whilst some principles and problems are discussed in depth and the bibliography provides access to an even more detailed treatment for the specialist, the primary aim has been to cover as many aspects of the subject as possible without excessive detail. Preference has been given to the more readily obtainable references whenever possible.

In the United States of America, the Department of Commerce has estimated that 280 kg of packaging materials go into the products used by each American every year. Packaging consumes about 80 per cent of all paper board produced, 65 per cent of all glass, 25 per cent of plastics, 22 per cent of paper, 15 per cent of wood and 7 per cent of steel. Obviously, this includes other areas as well as *food* packaging. However, it does underline the importance of packaging in the economy and the part played by materials other than plastics. Hence, whilst the major portion of the book is devoted to plastics, some consideration has also been given to the more traditional packaging materials such as glass, metals, ceramics and paper.

Most other books currently available have been written by specialists in the fields of plastics and polymer technology. It is hoped that a complementary view as seen by a food scientist and analytical chemist will be of some value.

N. T. Crosby

CONTENTS

ABBREVIATIONS

Å	Angstrom = 10^{-8} cm
A_w	Water activity
ABS	Acrylonitrile butadiene styrene polymer
ADI	Acceptable daily intake
AN	Acrylonitrile
BGA	Bundesgesundheitamt (Federal Health Office of the West German Federal Republic)
BHA	Butylated hydroxyanisole
BHT	Butylated hydroxytoluene
BIBRA	British Industrial Biological Research Association
BPF	British Plastics Federation
b.pt.	Boiling point
BSI	British Standards Institution
b.w.	Body weight
DMA	Dimethylacetamide
DMF	Dimethylformamide
DNA	Deoxyribonucleic acid
ECD	Electron capture detector
EEC	European Economic Community
EVA	Ethylene vinyl acetate copolymer
FAO	Food and Agriculture Organisation of the United Nations
FDA	Food and Drug Administration, USA
FID	Flame ionisation detector
GLC	Gas–liquid chromatography
GM	Global (or total) migration
GRP	Glass reinforced plastics

HDPE	High density polyethylene
HIPS	High impact polystyrene
HSGC	Headspace gas chromatography
i.d.	Internal diameter
IUPAC	International Union of Pure and Applied Chemistry
LDPE	Low density polyethylene
LD_{50}	Dose required to kill 50 per cent of a group of animals
MS	Mass spectrometry
OPP	Orientated polypropylene
PAM	Polyamide
PE	Polyethylene (polythene)
PET	Polyethyleneterephthalate
PIRA	Paper and Board, Printing and Packaging Industries, Research Association
PP	Polypropylene
ppb	Parts per billion $\equiv \mu g/kg$
ppm	Parts per million $\equiv mg/kg$
PS	Polystyrene
PTFE	Polytetrafluoroethylene
PVC	Polyvinylchloride
PVDC	Polyvinylidenechloride
RH	Relative humidity
SAN	Styrene acrylonitrile copolymer
SI	Statutory Instrument
STP	Standard temperature and pressure
THF	Tetrahydrofuran
TLV	Threshold limit value
TWA	Time weighted average
UHT	Ultra high temperature
VCM	Vinyl chloride monomer
VDC	Vinylidene chloride
WHO	World Health Organisation

SPECIFICATION OF CONCENTRATIONS

Trace amounts of additives and contaminants in food are usually specified in terms of parts per million (ppm) or parts per billion (ppb). In scientific and technical journals, however, the system of units known as SI (Système Internationale d'Unités) has now been adopted almost universally. SI is an extension and refinement of the traditional metric system, which is both logical and coherent and possesses many advantages in scientific work. Nevertheless, the above units are still widely used and, in general, are more readily understood. They have been adopted throughout this book. The relationship between the various units is shown below:

$$1\ \text{ppm} \equiv 1\ \text{mg/kg}$$

$$1\ \text{ppb} \equiv 1\ \mu\text{g/kg} \equiv 0{\cdot}001\ \text{ppm}$$

The concentration of vinyl chloride in air or gaseous samples is usually specified as a ratio of volumes rather than as w/v or w/w where:

$$1\ \text{ppm} \equiv 1\ \text{ml/m}^3$$

and

$$1\ \text{ppb} \equiv 1\ \mu\text{l/m}^3$$

Chapter 1

FOOD PACKAGING REQUIREMENTS

INTRODUCTION

From the earliest times Man has had to conserve and preserve his food so that an abundance obtained at harvest, or following a successful hunt, can be stored for use at a later date as a protection against famine until further supplies are obtained. Subsequent harvests, owing to adverse weather conditions, may be poorer in quality or quantity. Even today, the need to ensure an adequate supply of safe and nutritious food is a major problem for the developing countries throughout the world. Industrialised nations, such as those in the European Economic Community, have attempted to offset the vagaries of climate, which make food supplies so unpredictable, through a policy of financial intervention and market organisation designed to offer a fair standard of living to those working on the land as well as to ensure reasonable prices to consumers.

Few people would question the importance of good food, in adequate amounts, for the preservation of health. Wealthy, industrialised countries can 'make good' any shortages in supplies through imports. Currently, in the United Kingdom, about half of our food is imported. Hence, prime agricultural products worldwide have to be processed, or preserved, to a more stable form so that they can withstand transport, handling and storage over long periods of time. Very few foods, with the exception of some fruit and vegetables, etc., are consumed to any great extent in the raw state without cooking. Most foods are processed in some way, not only for reasons of preservation but also for palatability and to provide variety. Thus, basic ingredients as, for example, vegetable oils and starch can be used to prepare a wide range of manufactured products such as puddings, confectionery, desserts, etc. Changing social and work patterns in recent years have encouraged the growth and development of, so-called, 'convenience foods', with ramifications for the packaging industry as well. Hence, a food industry has evolved largely from the early nineteenth century onwards, when large numbers of the population began to move from the countryside into towns at the start of the industrial revolution.

TABLE 1
UK MARKET FOR PACKAGING MATERIALS BY VALUE (£ MILLION)

Packaging material	Year					Ratio
	1971	1973	1977	1978	1979	1979:1971
Tin plate	190	240	546	585	650	3·4
Fibreboard	151	205	423	492	585	3·9
Plastics (film plus containers)	100	165	448	450	746	7·5
Glass	94	114	283	316	354	3·8
Paper	85	108	182	198	226	2·7
Board	67	101	230	236	275	4·1
Aluminium foil	40	48	99	104	108	2·7
Cellulose (unconverted film value)	23	25	48·5	49	57	2·5

Abstracted from editions of *Packaging Review*, IPC Industrial Press Ltd, London.

Provision of food gradually developed into a specialised business as opposed to a personal or family operation hitherto. The formation of large armies to fight wars and, hence, the need for bulk food supplies also stimulated the growth of a food industry. In France around 1800, Appert showed that food could be preserved by heat treatment, without the use of chemicals, to destroy bacteria followed by a process of hermetical sealing to protect the food from subsequent contamination. For this work Appert was awarded a prize by Napoleon and from such origins the canning industry arose to the point where today, in the UK alone, some 6000 million cans of food are consumed annually. Other forms of packaging, e.g. glass, earthenware, wood, cloth and leather, have been used by primitive man from prehistoric times and some are still extensively used in modified form today (Table 1). These figures relate to packaging in general and cover the pharmaceutical and cosmetic industries as well as the food processing industry. In contrast, the use of various plastics for food packaging (Table 2) has only developed since around 1950, when there was a concomitant rapid rise in the number of food poisoning cases reported to

TABLE 2
UK CONSUMPTION OF PLASTICS IN PACKAGING
('000 TONS)

	1973	1979
LDPE	250	362
HDPE	75	134
PS	52	93
PP	90	79
PVC	56	61
PVDC	8	12
Polyester	—	3
Miscellaneous	5	7
Total	536	751

Abstracted from editions of *Packaging Review*, IPC Industrial Press Ltd, London.

the authorities (Table 3). Most probably this rise resulted from the growth in 'eating out' in restaurants, etc., following the end of World War II. The harmful effects of micro-organisms in foods have been described by Passmore.[1] The classical food poisoning micro-organisms are all bacteria, and following ingestion of contaminated foods, they produce symptoms of nausea, vomiting, diarrhoea or abdominal pain, depending on the organism. It has been estimated that 45 per cent of all meat and 60 per cent of poultry are contaminated by *Salmonella* spp. but they are heat sensitive and so destroyed on cooking. These organisms, which are responsible for

TABLE 3
REPORTED CASES OF FOOD POISONING IN THE UK

Year	Number of cases reported
1940	47
1944	275
1948	964
1955	8 961
1965	8 313
1970	8 088
1975	10 936
1977	10 365

From Annual Reports of the Chief Medical Officer of the Department of Health and Social Security, HMSO, London.

around two-thirds of all notified cases of food poisoning, can survive for long periods in frozen foods. Hence, poultry must be completely thawed before cooking to ensure complete destruction of all organisms at the centre of the carcass. Strict attention to hygienic handling and storage of fresh and cooked foods will prevent problems associated with food poisoning, both in the home and at mass catering establishments; so reducing the continuing upward trend in bacterial food poisoning (see Table 3). Undoubtedly, such cases represent only a fraction of the total number of incidents occurring in any one-year, particularly when only mild symptoms are involved. This need for better hygiene, together with the introduction of pre-packaging of foodstuffs, resulted in new outlets for plastics and a rapidly developing food packaging industry. Tables 1 and 2 show how, even in the 1970s, the growth in plastics for packaging has outstripped growth in other materials, despite economic pressures resulting from the change in crude oil prices during this period. Moreover, consideration of packaging materials consumption by volume instead of by value, shows that there has been little real growth except in the plastics sector.

Whilst packaging is required to fulfil a number of different functions, its primary role is to retard or prevent loss of quality (nutritional and aesthetic) and to give protection against environmental contamination. Some of these spoilage mechanisms and the measures taken to control their effects will now be considered in greater detail, since they influence the selection of the best material for packaging.

Food Spoilage Mechanisms

During growth, transport, processing and storage, food may deteriorate in quality or be lost in quantity through one or more of the following mechanisms:

(1) Depredation by macro-organisms, e.g. pests and vermin.
(2) Contamination by growth of micro-organisms, e.g. bacteria, fungi and yeasts.
(3) Chemical reaction brought about by enzymes, through oxidation or hydrolysis.
(4) Physical changes, e.g. loss of moisture with associated texture effects, ingress of gases, taint, radiation and mass transfer.

Pests such as weevils and other beetles, ants, flies, cockroaches, or worms may contaminate growing as well as stored crops. Strict quality control of raw materials is essential therefore to prevent contamination of the finished product. Many species of insects can develop at water activity (A_w) levels below 0·3, but multiplication is prevented at temperatures below 15 °C. Vermin (such as rats and mice) consume the crop, and sometimes the packaging material, as well as (cf. insects) causing contamination by mechanical transmission of disease-producing organisms as they move from excrement to food, thereby resulting in the destruction and condemnation of the total consignment. The presence of micro-organisms in foods, whilst in many cases[1] producing disease, sickness, possible loss of life and further wastage, can also be beneficial. Fermentation processes and the ripening of cheese are examples of beneficial changes occurring as a result of microbiological inoculation. Chemical changes are mainly undesirable in that hydrolysis or oxidation often lead to the production of off-flavours, e.g. rancidity, although many alcoholic beverages improve on keeping through chemical reactions such as esterification. Enzymes, proteinaceous compounds naturally occurring in the tissues of plants and animals, continue to act after the harvesting of crops or the death of the animal. In some cases these changes are undesirable, e.g. the browning action in apples. Hence, vegetables are usually blanched prior to freezer storage. Alternatively, some enzymatic changes are encouraged, e.g. the hanging of game to develop flavour before cooking and eating. Physical changes involve principally a loss of moisture with associated textural changes. The absorption of volatile compounds may produce problems of taint, although the useful storage life of some foods may be prolonged in atmospheres of carbon dioxide or nitrogen.

Micro-organisms, including moulds, yeasts and bacteria, develop rapidly under favourable environmental conditions. Hence, most control mechanisms are designed to change these conditions so as to make them less favourable to growth.

Control Mechanisms

Whilst some of the above agents produce desirable changes in the organoleptic qualities of a food, the majority lead to loss of appearance, taste, nutritional quality and safety. Hence, a number of preventive measures have been devised over the years to control or retard loss of crops and food products, and any associated deterioration in quality. These measures can be grouped as follows:

(1) Reduction of moisture; drying, salting.
(2) Chemical preservation using sulphur dioxide, nitrate/nitrite, pesticides, organic acids and their derivatives, acidulants (pickling), antioxidants.

(3) Use of high temperatures; canning, pasteurisation.
(4) Use of low temperatures; refrigeration, freezing.
(5) Irradiation.
(6) Packaging, for protection and for a controlled environment.

Since all forms of life require water for growth and even survival, dehydration of food is an effective means of restricting deterioration caused by fungi and other micro-organisms. The reduction in water activity can be effected by simple drying, or by the use of sugar (as in jams), or salt (through osmotic effects), or by smoking. In the latter case, whilst the main change is a reduction in moisture content, some chemical action through the presence of phenolic compounds and formaldehyde in the smoke, may also occur. A reduction in moisture content may also produce important savings in transport costs. Chemical preservatives comprise a wide and heterogeneous group of products, from insecticides and fungicides used in agriculture, to simple inorganic and organic preparations added during food manufacture to kill-off various groups of micro-organisms, or to restrict their growth. A more detailed account of the science and technology of food preservation can be found in a recent review.[2] Most organisms grow rapidly at human body temperature (37 °C). Hence, a change in temperature, away from this optimum, should prevent or reduce their rate of growth by the creation of an unfavourable environment. High temperatures may effect a complete sterilisation or, alternatively, only partial stabilisation may be obtained. Usually the heat treatment is sufficient to kill any pathogenic organisms, fungal spores and to inactivate enzymes by denaturation. The heat treatment itself may lead to some loss of food quality. Hence, only the minimum necessary to safeguard the product is carried out. In the case of milk there are a number of heat treatment processes designed to kill pathogenic organisms only (pasteurisation) or all micro-organisms (sterilisation). Milk in the cow's udder is probably sterile but it readily picks up contaminating organisms in passing through the teat and into the milking pail or from other dairy equipment. Milk becomes sour in a few hours unless kept cool. The pathogenic organisms present in milk are all heat sensitive but there is a non-linear relationship between the temperature and time of heating required for safety. Some examples of suitable conditions for pasteurisation are given below:

40 min at 57 °C
10 s at 71 °C
$<$2 s at 75 °C

As the temperature is increased, less time is required. However, the time needed falls much more rapidly than would be expected on a linear basis. High temperatures tend to give the milk a slightly cooked flavour. Fifteen seconds heat treatment at 73 °C is widely used in many countries for the pasteurisation of milk. Unfortunately this leads to some loss of nutritional value. Exposure to light (especially in the UK where milk is extensively packed in glass bottles and exposed to direct sunlight during delivery and on the doorstep) leads to a high loss of vitamin C and over 50 per cent of riboflavin (vitamin B_2) as well as a change in the flavour. Vitamin C is converted to dehydroascorbic acid which, whilst still biologically active, is very unstable to heat and is lost during pasteurisation or heat treatment in the home. Removal of oxygen from milk (deaeration) before heat treatment reduces the loss of vitamin C. Some thiamine (vitamin B_1) and vitamin B_{12} are also lost during pasteurisation.

Sterilised milk has an almost unlimited keeping quality. A typical process involves pre-sterilisation at 120–135 °C for a few seconds, homogenisation to prevent separation of fat before filling at 60–70 °C into hot glass bottles which are hermetically sealed and heated to ensure that the centre reaches a temperature of 115 °C for 5 min. Some 30 per cent of the total content of thiamine is lost as well as 50 per cent of the vitamin C and nearly all the vitamin B_{12}. Vitamin A and riboflavin are not affected, except by light.

UHT (ultra high temperature) milk is usually packed in cartons. The milk is heated in a continuous process to 130–150 °C and held for 2–4 s using plate or tube heat exchangers. Homogenisation is also necessary. Often plastic-coated, aluminium-lined containers are used to protect milk from exposure to light. Vitamin C and B_{12} losses are minimal if the milk is deaerated. The product is equal in flavour, nutritional value and colour to pasteurised milk but has vastly superior keeping qualities.

Freezing

Eskimos have used low temperatures for generations, for the storage and preservation of food. However, the frozen food industry developed only from 1930 onwards, following the invention of the quick freezing process developed by Birds Eye, which reduces the size of ice crystals formed. In turn, this reduces damage to cell and tissue structure and drip loss on subsequent thawing. The zone of maximum crystallisation is 0 to −5 °C and modern methods of freezing are designed to pass through this range as rapidly as possible. The use of very cold temperatures (e.g. liquid nitrogen, −200 °C) can produce undesirable dimensional stress. Nevertheless, modern methods of freezing have made possible the preservation of a wide range of foods, with little or no loss of natural flavour, colour or nutritional quality. They have also made a significant contribution to the growth in the popularity of convenience foods and the frozen food industry is expected to grow by another 20 per cent by 1987.

Whilst some species of micro-organisms are reduced in number on freezing, others are preserved and can multiply after thawing. However, no known pathogenic organism grows at temperatures below 4 °C and virtually all microbial activity ceases at −10 °C. Chemical reactions are slowed down as the temperature falls until at −18 °C (0 °F) most foods can be stored safely for extended periods. Domestic refrigerators possess an average temperature of about 5 to 10 °C in the main cabinet depending on use. The temperature in the frozen food compartment depends on the 'star rating'. A one star * appliance will reach −6 °C, two star ** −12 °C and three star *** −18 °C. Freezers are normally maintained at −18 °C but include a fast-freeze switch permitting lower temperatures which prevents damage to the contents on the introduction of substantial amounts of unfrozen food. Foods are blast-frozen commercially at around −34 °C (−30 °F) by high pressure fans at speeds which cannot be matched by home freezers. Even lower temperatures are used in cold stores (down to −40 °C; by a mathematical coincidence this temperature is also equivalent to −40 ° on the Fahrenheit scale) for long-term storage, although bacon and other high electrolyte foods are an exception in that they store less well at such temperatures.[3] Fluctuations in the freezer temperature during storage cause nucleation of ice crystals and fogging and consequently, are to be avoided. In order to inactivate enzymes, vegetables are 'blanched' by immersion in boiling water for a few minutes prior to freezing. Otherwise, off-flavours, loss of colour and texture may result on prolonged storage. For short periods, blanching may be unnecessary. Fruits cannot be blanched and are usually packed with sugar or in a syrup.

The main packaging requirements of materials for frozen foods are the ability to withstand low temperatures and conversely, in some cases, temperatures up to 100 °C e.g. boilable bags. The material must also prevent desiccation and the transfer of oxygen, flavour volatiles, taint or grease. Shrink wrapping is widely practised for products of irregular shape e.g., poultry, in order to exclude as much oxygen as possible.

Irradiation

Irradiation techniques have the advantage that sterilisation can be effected after packaging, thus eliminating insects and hazardous bacteria such as *Salmonella*. The technique can also be used to delay the maturation of fruit and vegetables. However, penetration can sometimes be a problem.

Furthermore, little is known about the nature of breakdown products and off-flavours are sometimes produced. Sterilisation by irradiation is not currently permitted in the UK.

FUNCTIONS OF PACKAGING

The role of packaging in food preservation is essentially one of protection against extraneous agents, and dirty handling. These include all the spoilage agents mentioned previously. Obviously if food is already contaminated before packing, the latter can have little preventive effect. However, most foods are subjected to some form of preservation before packing. The packing material may, however, further extend the shelf life of the product. Furthermore, packaging is used to present the commodity in an attractive form to the buyer.

Before describing the properties of individual plastics, it is pertinent to consider the requirements of an ideal material for the packaging of foodstuffs. Packaging embraces both the art and science of preparing goods for storage, transport and eventually sale. Hence, packaging has other functions in addition to the purely protective aspect outlined in the introduction. The successful application of plastics in the field of packaging may, therefore, require the collaboration of the food manufacturer, the plastics producer and converter, the designer and ultimately the consumer through market research activities. Economic considerations impose further restrictions, in that consumers are interested only in the product and not its package. Hence, the package should be as simple and as inexpensive as possible, whilst remaining consistent with the primary objectives of protection and attractiveness at the point of sale. Further constraints may be imposed by the need to use machinery during production processes, so that qualities such as flexibility, printability, ease of working on wrapping machines, heat-sealability and suitability for forming by blow-moulding, extrusion, vacuum or thermal techniques may also influence initially the material to be used. Other desirable properties of particular importance in the food industry include transparency and permeability (or non-permeability depending on the product) to water vapour and to gases such as carbon dioxide, oxygen or nitrogen.

Problems Encountered in Food Packaging

Whilst many technological problems encountered in the use of plastics for food packaging applications are those that are common to the packaging industry as a whole, some other considerations arise solely through the nature of the product to be packaged. Most foods are perishable, having a useful life which can be as short as only a day or two in the case of some fresh foods, depending on the conditions of storage. Hence, requirements for a food packaging material are generally more stringent than for many other retail and household commodities.

The principal intrinsic requirements of a *food* packaging material can be classified as follows:

(1) Transparency and surface gloss for customer appeal.
(2) Control over transfer of moisture.
(3) Control over transfer of other gases/vapours.
(4) Wide temperature range in storage and use.
(5) No toxic constituents.
(6) Low cost.
(7) Protection against crushing.

Customers like to see what they are buying and, hence, transparency (with gloss) is an essential requirement in many food packaging applications. Opaque materials such as trays, although widely used, prevent the customer seeing both sides of the product and, in the case of a joint of meat, this can engender distrust and consequently sales resistance. Control over changes in the moisture content of a food during storage is most important. Some products, e.g. snack foods, biscuits, or boiled sweets, must be packed in a material with a very low permeability to water vapour to preserve crispness. It is also essential to reduce moisture losses with many foods and again a low permeability film would be required. In other cases, some loss of moisture is desirable to avoid sweating, condensation on the inside of the package and concomitant loss of transparency as well as increased risk of mould growth.

In addition to transfer of water vapour in and out of the packaging material, control over permeability to other gases such as oxygen and carbon dioxide is also very important.

Most fresh foods need to 'breathe', hence the packaging material used for these products must allow ingress of oxygen and respiration of carbon dioxide. Where the material chosen does not permit sufficient gas transfer, the problem can often be solved by incorporation of a few holes punched into the film. Fresh meat also requires ingress of oxygen to maintain a satisfactory surface colour. On the other hand, foods with a high fat content (dairy products, bacon, crisps, etc.) become rancid on exposure to oxygen and are often vacuum packed, or packaged in an inert atmosphere, using a material of very low permeability. Low permeability materials are also useful for the packaging of fish, or coffee, where the odour must be contained strictly within the package. Foods subject to atmospheric oxidation are often protected by the addition of an antioxidant, e.g. BHA or BHT. Some products are delicate, e.g. cereals and crisps, and need to be loosely packed. This increases the air inside the package and, coupled with the high surface area of such products, can lead to rancidity. Additional protection, and longer storage life, can be obtained by addition of antioxidant to the packaging material itself as well as to the food. This reduces the concentration of antioxidant required in the food and since only the surface of the food needs protection most added antioxidant is therefore wasted inside the body of the product. Furthermore, customers are likely to prefer addition of preservatives to the packaging material rather than to the food itself, although migration of the antioxidant from the package to the food may still occur.

Packaging of Selected Food Groups

The application of the general principles of food packaging outlined above to specific food categories will now be discussed, indicating the packaging requirements imposed by both the properties of the food as well as the marketing needs. Materials commonly used in response to these needs are indicated, although wide variations occur throughout the world as dictated by local factors. Furthermore, the industry is in a constant state of change with new products being introduced continually.

Bakery and Confectionery Goods

This group includes bread, rolls, cakes, pastries, pies, biscuits and crackers. The principal problem affecting soft, baked goods is staling, which packaging can do little to arrest. Since the staling process takes place so rapidly, bakeries are located close to the point of consumption and produce daily. Numerous small independent firms flourish, in addition to the large multiple businesses. During staling, hydrated starch reverts to the crystalline form, i.e. the swollen granules change to a less tender form. The low fat content and short shelf life of these products renders protection against oxidative deterioration unnecessary. Indeed, bread is often sold without any wrapping at all. Thin film or paper to protect against handling is often all that is required, although polythene bags have been used particularly for sliced loaves and for moisture retention for bread that is to be frozen. These products are soft and elastic, hence some protection such as racking is required to prevent crushing during stacking in delivery vans, etc. Sweet, baked goods, e.g. doughnuts and cakes have a much higher fat and sugar content, which retards staling and helps to reduce microbiological deterioration. A greater greaseproofness is required and cakes are even more subject to damage by crushing than bread. Cakes are often packed in paperboard bags, or rigid cartons, often with transparent cellophane windows or overwrapping. Pastry products are packed in cling film, or plastic nests, or aluminium foil base plates; again, a good fat-resistant wrap is required. Crackers, biscuits and wafers have a much longer shelf life owing to their low moisture content. The wrapping must provide an excellent moisture barrier to prevent loss of crispness and texture. Increased fat content of these products imparts a shorter shelf life but the wrapper must act as an efficient grease barrier. The fats used are normally hydrogenated and protected with an antioxidant. Cellophane, waxed glassine and biaxially orientated polypropylene have all been used for this type of application.

Cereals

Ready-to-eat breakfast cereals need to remain crisp. They possess a long shelf life if not opened and are, therefore, manufactured centrally and widely distributed. The main need is protection against moisture pick-up, loss of crispness and toughening texture. This process is, however, reversible. They are packaged in waxed paper inside paperboard cartons to give some protection against crumbling through pressure on stacking and transport. Some cereals are packeted in individual portions using polythene-coated paper. The PE acts both as a moisture barrier and heat sealant. Where protection against surface oxidation is required, an antioxidant (usually BHA or BHT) is added to the paraffin wax used on the glassine liner as well as in the foodstuff for maximum protection.

Cereals to be eaten hot are processed to reduce cooking time in the home by pre-cooking or enzyme treatment. This hydrates the starch granules as in confectionery products. The fatty fraction of the grain is frequently removed. Hence, there is little danger of fat exudation or rancidity. As the product is dry there is little danger from microbial deterioration. They are packed simply in lined paperboard cartons. Sweetened products need some protection against moisture pick-up which produces stickiness. Generally, packaging requirements are less exacting although insect infestation can be a problem in improperly closed containers.

Chocolate and Sweets

Chocolate has a high fat content and needs to be protected against fat seepage. Chocolate bars are wrapped in aluminium foil or glassine with a sleeve overwrap. Filled chocolates have individual glassine cups or PVC trays and are packed in boxes overwrapped in transparent film, e.g. cellophane or polypropylene to protect them from dirt and dust. These boxes may be stored for several months before sale. Temperature fluctuations or excessive loss of moisture can cause surface fat bloom, otherwise packaging requirements are not too critical apart from resistance to grease. However, opaque materials are preferred since fat seepage is thereby masked.

Sweets and jellies are packed in plastic bags or cardboard tubes or wrapped in aluminium foil/paper laminates. Some sweets are individually wrapped in cellophane, polypropylene or waxed paper twists. Moisture loss must be prevented and pick-up of water vapour also causes problems in these products with a high sugar content. Caramels become brittle or sticky. Loss of moisture from marshmallows and jellies can lead to sugar crystallisation and hardening.

Coffee

Products currently on the market include roasted and ground coffee for percolating as well as spray- or freeze-dried instant coffees. Coffee is consumed primarily for its flavour, which is developed during roasting. It is packed in hermetically sealed cans, under vacuum, to exclude oxygen. Once opened, the product rapidly loses its aroma by volatilisation and coffee oils are oxidised on exposure to air. Tins are provided with polythene lids for re-closure. Instant coffee is a hot water extract that has been spray-, freeze- or roller-dried and is packed almost exclusively in glass jars with waxed paper seals. Dried coffee is hygroscopic but less subject to atmospheric oxidation as little oil is extracted compared to coffee beans. Freeze-dried products are even more hygroscopic and contain more flavour volatiles since they have not been subjected to the hot extraction process. Some coffee is packed in polyester/aluminium/polythene pouches.

Convenience Foods

These comprise individual portions of sauces, spices, sugar, jams or cream formerly dispensed from bulk containers, and are principally used in catering establishments. Paper coated with polythene for heat sealability and moisture resistance is frequently used. Jams and jellies are packed in plastic tubs or tubes with foil peel-off lids. Such products with a high sugar. content are subject to mould growth and have to be hot filled, hence, the pack must be heat resistant. Clear polystyrene tubs are suitable.

Condiments are packed in laminated envelope pouches or plastic or metal tubes.

Dairy Products

This group includes milk, butter, margarine (because of its traditional relationship with butter), ice cream, yoghurt and cheese. These products are characterised by a high fat and water content. For milk, glass is still the major packaging material used in the UK although waxed or coated paper cartons are also used to a limited extent. Ice cream is filled in the partially frozen state, then frozen. The package has to withstand temperatures down to $-35\,°C$ ($-30\,°F$). Materials used include rigid waxed/ethylene vinyl acetate copolymer-coated paperboard cartons, or polystyrene containers. Yoghurt is often packed in similar containers; acid resistance and flavour impermeability being the major requirements. Cream cheese is packed in aluminium foil laminates, whilst cottage cheese is sold in waxed paperboard or polystyrene tubs and lids. Sliced, processed cheese is packed under a carbon dioxide flush in coated cellophane; slices being interleaved to aid separation. Butter and margarine have much higher fat and lower moisture contents and need to be refrigerated to maintain flavour. Paper, aluminium foil and plastic tubs are all used for packaging.

Dehydrated Foods

This group comprises products as varied as soups, mashed potato, fruits, bakery mixes, pasta products and rice. The principal requirement is obviously the exclusion of water vapour. Oxygen exclusion may also be important in some cases, e.g. for vegetables and soups, and gas flushing is used for products such as dried milk and egg powders which are even more sensitive to oxidation. Laminates incorporating aluminium as a barrier are best, e.g. polyethylene–aluminium–paper. Dried fruits have higher A_w activity and can be stored safely in permeable film.

Frozen Foods

These are packed to retard moisture loss (by sublimation and associated freezer burn), loss of volatile aromatics, and to exclude oxygen and light. Freezer burn (severe desiccation of the surface layers of the flesh and oxidative deterioration) is caused by temperature fluctuation and moisture migration. Hence, headspace volume should be reduced as much as possible by using shrinkable films. The effect of low temperatures is the major influence on packaging requirements. Polyethylene bags are widely used along with aluminium foil and dishes which will withstand very low temperatures without embrittlement or shock fracture.

Meat

Fresh meat is packed to protect against external dirt and handling, to retard moisture loss and to maintain a bright cherry red surface colour. For this, ingress of oxygen to convert the purplish muscle tissue to bright red oxyhaemoglobin, is essential; PVC and cellophane films are used for this purpose. For longer term storage (>2 to 3 days), films impermeable to oxygen and water vapour are required. Copolymers of VDC and VC are used since they are suitable for vacuum packaging. On exposure to air the meat changes from a dark colour to red. In processed meat products the colour is stabilised by curing agents such as nitrite. Such products are usually packed in an oxygen-free environment with a good gas barrier material such as PVDC to protect the pigment from oxidation, or canned for longer term storage.

Snacks

These include, potato crisps, peanuts, etc. They have a high fat content and need protection both against moisture ingress and loss of crispness, and oxidative rancidity despite the use of antioxidants. For this purpose coated films e.g. PVDC on glassine, cellulose or polypropylene, are widely used. Moisture levels following manufacture are usually as low as 2 per cent. An increase to only 4 per cent would result in loss of crispness and texture. For long-term storage, vacuum packaging in cans or glass jars is practised.

Wine and Other Beverages

Good quality wine has always been (and probably always will be) packed in glass bottles with a cork seal. Such a container permits only tiny amounts of oxygen to enter via the cork and so react with the contents. Recently, however, experiments have been carried out using plastic packaging for 'vin ordinaire', as well as for draught beer, cider and carbonated soft drinks. Bag-in-box containers made of LDPE with a wall thickness of around 1 mm have been used with capacities in the range 1 to 5 gallons. Unfortunately, in contrast to the glass bottle, plastics are not impermeable. LDPE in particular is highly permeable to oxygen and similarly allows flavour volatiles to escape, so that these packages are suitable only for short-term storage. Permeation by oxygen destroys residual sulphur dioxide added as a preservative, leading subsequently to the development of off-flavours. In addition, alcohol is an excellent extraction solvent for additives in the plastic, which will migrate into the wine and affect its organoleptic qualities. Some newer containers are currently undergoing trials. Plastics such as PET, PVC, SAN and some composite materials have a much lower permeability than LDPE. PET bottles are already being used for soft drinks where impermeability is required to prevent the loss of carbon dioxide. Injection moulded EVA co-polymer 'corks' for wine bottles are now available. They are porous and rubbery with a smooth finish. Unlike the natural material they do not have to be kept moist to prevent shrinkage and leakage. Bottles can, therefore, be stored upright.

CONCLUSION

Within the above technical limitations, price restrictions will be the dominating factor governing selection of a material for packaging, particularly in an industry such as the food industry, where price competition is so keen. There is a growing consumer resistance to wasteful packaging, especially for basic items such as food. Elaborate and expensive outer packaging whilst acceptable for luxury items in the cosmetic field will become increasingly unacceptable for basic necessities, where the packaging is simply functional and consumer interest is concerned solely with the product rather than its wrapping.

Food Packaging Materials

The following materials are widely used for food packaging, or in food contact applications:

(1) Paper and coated-paper products.
(2) Cellulose products, cellophane.
(3) Metals—tinplate, aluminium, stainless steel, pewter.
(4) Ceramics.
(5) Glass.
(6) Rubber.
(7) Plastics.
(8) Miscellaneous—wood, fabric, etc.

Most of these materials have been in use for many years and have given rise to very few problems. The old practice of shipping tea in lead-lined chests has now been discontinued and equally the use of improved glazes on ceramic ware has reduced contamination by metals such as lead and cadmium. Plastics, on the other hand, whilst offering many advantages as a new range of packaging materials, at the same time involve contact between food and a whole new range of chemical components not previously used in the food industry and for which no previous experience was available. This problem was of particular concern since packaging can involve a long and intimate contact between the food and its container during storage at wholesale and retail outlets and in the home. In addition to the polymers used in the manufacture of food packaging materials, a large number of chemical adjuncts may also be incorporated into the finished product. Such agents include plasticisers, antioxidants, release compounds, heat and light stabilisers, lubricants, antistatic chemicals, adhesives, pigments, and many other compounds, which are added by the manufacturer to alter the processing, physical properties, or stability of the final product. The addition of such substances is essential to achieve the desired chemical, physical, and mechanical properties in the finished article. However, the use of such a wide range of chemicals inevitably gives rise to concern amongst both legislators and consumers. The problem was intensified with the discovery of toxic and carcinogenic properties in some of the compounds used. This stimulated whole new areas of research into the toxicology and epidemiology of such compounds, as well as analytical studies to develop exceptionally sensitive methods of analysis to measure residual levels of monomers in packaging materials, and in the diverse types of food which comprise the national diet. Residual levels of such substances are not a cause for concern unless they are transferred from the package into the foodstuff. Hence, in addition to analytical studies, a more fundamental study of the mechanisms of this transfer (migration) has been undertaken.

The chief advantage in using plastics for packaging purposes is that most polymers possess excellent physical properties such as strength and toughness, combined with low weight and flexibility, as well as resistance to cracking.

As a consequence of the diversity of packaging applications in the food industry, a wide range of polymer materials has been developed over the years. The basic material (with additives as necessary) is then converted into film, powder or sheet and moulded, or further processed into containers, e.g. trays, tubs, bags, pouches, sachets, blister packs or shrink wraps. Frequently, no one material possesses all the desirable properties required and, hence, copolymers, or even laminated materials consisting of two or more layers of different polymers (having different properties) cemented together, may have to be used.

With the above miscellany and multiplicity of requirements in mind, the major types of polymer used in the food packaging industry will now be surveyed.

REFERENCES

1. PASSMORE, S., Bacterial food poisoning, *Nutrit. Food Sci.*, 1979, **56,** 7.
2. TILBURY, R. H. (Ed.), *Developments in food preservatives*, Applied Science Publishers, London, 1980.
3. SYMONS, H. W., Frozen foods and their stability, *Proc. Inst. Food Sci. Technol.*, 1977, **10,** 153.

SUGGESTIONS FOR FURTHER READING

PAINE, F. A. (Ed.), *Fundamentals of packaging*, Blackie, London, 1962.
SACHAROW, S. and GRIFFIN, R. C., *Food packaging*, AVI, Westport, Connecticut, 1970.

Chapter 2

PLASTIC PACKAGING MATERIALS

INTRODUCTION

The principal polymers found in the food industry are listed below, with a brief outline of the major manufacturing processes used and the distinctive properties of the final product. Further details of the technology of polymer chemistry and plastics manufacture can be obtained from the bibliography at the end of the chapter. A summary of some physical properties of the main polymer types is presented in Table 1. For each product, a broad spectrum of different compositions can be manufactured and, hence, the resulting properties also vary over a wide range. However, some indication of the differences between the different types of polymer can be seen by perusal of the table. The importance of the permeability to oxygen and water vapour to its suitability as a food packaging material has been stressed in Chapter 1. Hence, Table 2 shows the relative permeability of some polymer films to both oxygen and water vapour. The properties of individual monomers can be found in Chapter 3, Table 1.

Most polymers used for food packaging are thermoplastics, i.e. they can be softened by heating and hardened on cooling, repeatedly, provided that no chemical decomposition occurs. Thermosets, on the other hand, undergo an irreversible change on heating and do not soften. They char at high temperatures since their molecular structure is a complex three-dimensional bonded network.

Thermoplastics are preferred since the basic polymer may have to be subjected to several heating and cooling cycles following synthesis and during manufacture into formed containers or film. Thermosets could not be processed in this way. The main use of thermosets in the food industry is

TABLE 1

SOME PHYSICAL PROPERTIES OF PLASTICS USED IN FOOD CONTACT APPLICATIONS

Property	LDPE	HDPE	PP	R-PVC[a]	PVDC	PTFE	HI-PS[b]	HI-ABS[b]	Nylon II	Polycarb
Molecular weight $\bar{M}n$	1–3 × 10^4	10^5 +	75–200 × 10^3	5–12 × 10^4	—	4 × 10^5–10^6	10^4–10^6	—	—	≤2 × 10^5
$\bar{M}w$	~3 × 10^5	~1·25 × 10^5	3–7 × 10^5	—	—	—	—	—	—	—
Crystallinity, per cent	55–70	80–95	65–70	<5	—	93–98	—	—	—	—
Density, g/cm^3	0·915–0·935	0·945–0·965	0·90	1·4	1·68–1·75	2·13–2·2	1·04–1·11	1·02–1·05	1·04	1·20
Softening point, °C	86	120–130	150	70–80	100–150	—	78–103	85	175	135–165
Melting point, °C	112	137	168	220	—	342	240	—	185	220–230
Tensile strength, lb/sq in × 10^{-3}	0·8–2	2·4–3·2	100–600	5–9	5–10	2·5–5·0	~400	8·8	7·8	8–9
kg/m^2 × 10^{-3}	0·06–0·14	0·17–0·22	7·0–42·0	0·35–0·63	0·35–0·70	0·17–0·35	~28	0·62	0·55	0·56–0·63
Elongation at break, per cent	—	20	—	2–40	15–25	200–400	<3	20	2–300	5–7
Refractive index, n_D^{25}	1·51	1·54	1·49	1·52–1·55	1·61	1·35	1·59–1·6	—	1·52	1·585
Glass transition temperature (T_g), °C	−20	−125	−18	80–85	—	—	100	—	—	—
Water absorption in 24 h, per cent	0–75	0–40	<0·01–0·03	0·2–1·0	>0·1	0·01	0·05–0·22	0·2–0·45	0·4	0·15

[a] Unplasticised.
[b] High impact.

TABLE 2
OXYGEN AND WATER VAPOUR PERMEABILITY OF SOME POLYMER FILMS

Film (0·001 in thickness)	Permeability [cm^3 (STP)/m^2/24 h/cm Hg at 25 °C, 65 % RH]	
	Oxygen	Water vapour
LDPE	120	3 200
PP	52	1 500
PVC	39	5 300
Nylon II	4·6	4 700
PET	1·1	4 400
PVDC	0·11	160

From: Davis, E. G., *Food Technol. Aust.*, 1970, **22,** 62.

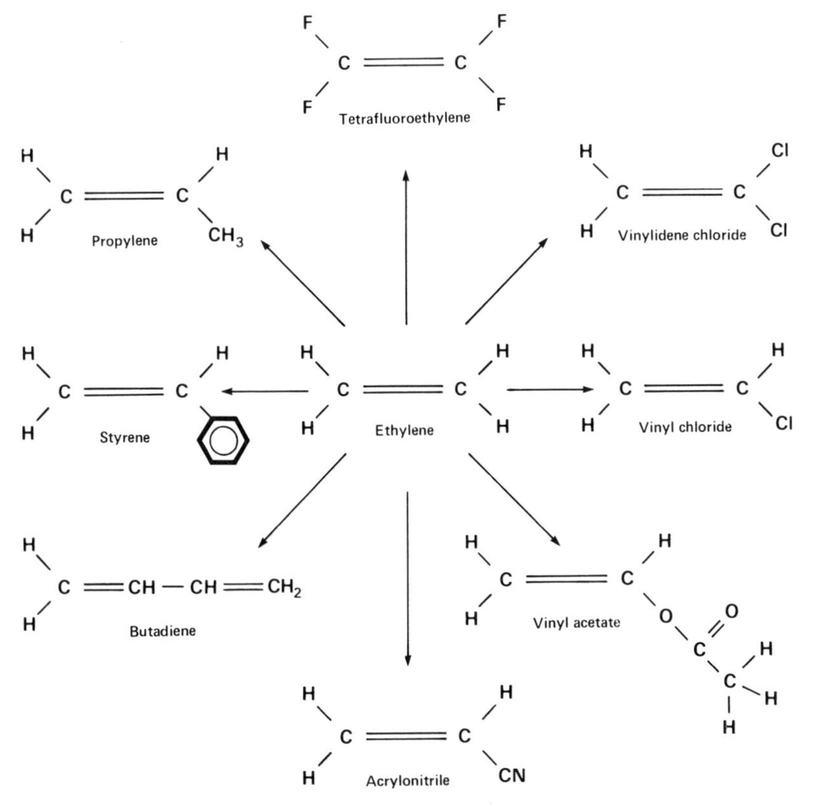

FIG. 1. Family of ethylenic thermoplastics.

as bottle caps, and in most cases direct contact with the food does not occur because of the use of liners as a barrier.

Most thermoplastics can be regarded chemically as derivatives of ethylene ($CH_2{=}CH_2$). The relationship between the various monomers used in the polymerisation processes is shown in Fig. 1. They are also often referred to as the vinyl plastics since they all contain the vinyl grouping ($CH_2{=}CH{-}$), or polyolefins since the monomers all contain an olefin linkage before polymerisation.

POLYOLEFINS

Polyethylene (or Polythene)

Polythene (PE) is made by the addition polymerisation of ethylene gas obtained as a by-product of the coal and oil industries. Two main processes have been described giving rise to two different products, although materials intermediate in composition between these extremes are also available. Polymerisation under high pressure (typically 1000–3000 atm), produces a macromolecule with a high degree of branching i.e. a mixture of long single chains and chains with branches linked at right angles to the main chain (see diagram below), whereas the low pressure process favours the formation of long parallel linear chains. The molecular formula of the polymer is $(CH_2)_n$ and, although the chains in the macromolecule are referred to as linear, in fact, the carbon atoms zig-zag as shown below:

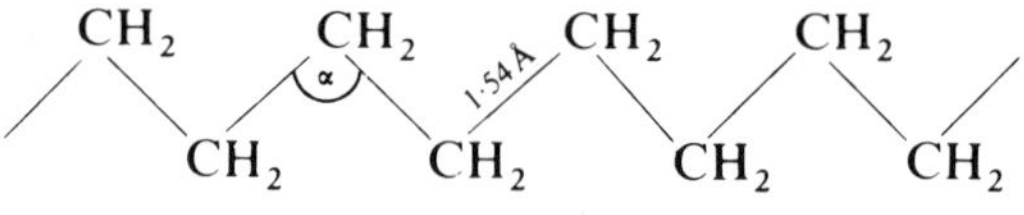

CH_2 CH_2 (α) 1·54 Å CH_2 CH_2

CH_2 CH_2 CH_2 CH_2

Linear chains in the polythene macromolecule

The carbon atoms are linked by strong covalent bonds at a distance of 1·54 Å (15·4 nm) at an angle (α) = 109°. In addition, neighbouring chains are stabilised by much weaker van der Waals' forces which account for much of the 'plasticity' observed with this type of material. Although individually the secondary bonding forces are very weak, over the whole length of the chain they can exert a most important influence on a number of physical properties.

The presence of branched chains prevents close packing of the polymer chains and, hence, the density of the finished product is low. It is

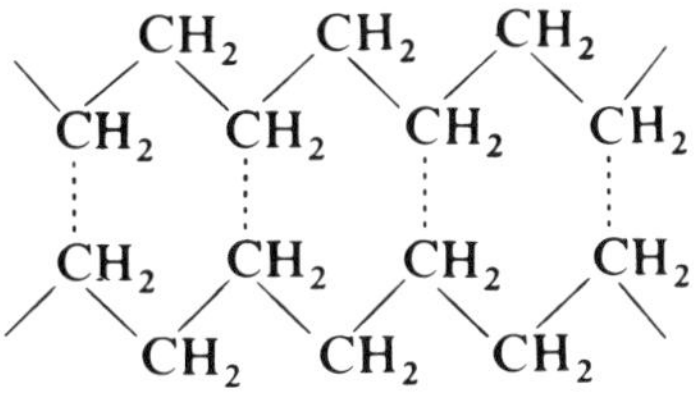

CH_2 CH_2 CH_2

CH_2 CH_2 CH_2 CH_2

CH_2 CH_2 CH_2 CH_2

CH_2 CH_2 CH_2

Secondary forces between linear chains

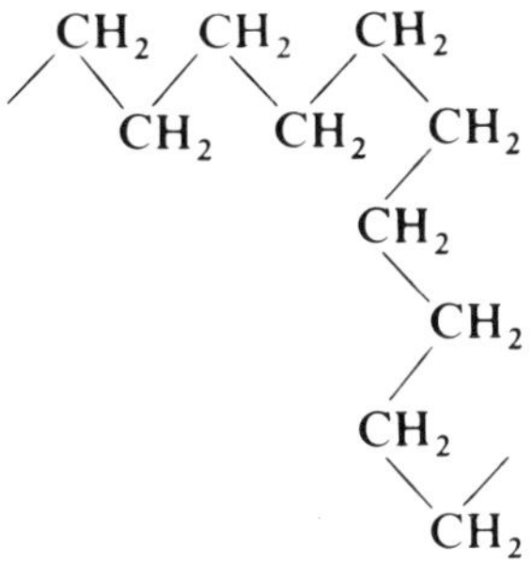

CH_2 CH_2 CH_2

CH_2 CH_2 CH_2

CH_2

CH_2

CH_2

CH_2

Branched chain polymer

unfortunate and, naturally, the subject of some confusion, that low density polyethylene (LDPE) as the product is known, should be produced by a high pressure process. The irregularity introduced by the presence of many branched chains also lowers the overall degree of crystallinity (regularly ordered regions) and the softening point, since less energy in the form of heat is then required to break the secondary bonding forces between the chains which are less closely and regularly packed. Crystalline regions of the polymer consist of parallel chains of linked monomer molecules, whilst amorphous regions are randomly orientated chains irregular in configuration. Crystalline regions are not as regularly ordered as inorganic crystalline materials and there are gradual transitions from crystalline to wholly amorphous regions within the polymer macromolecule. Highly symmetrical molecules without branched chains or bulky side chains produce the most highly ordered polymers.

LDPE is a tough, slightly translucent, flexible material with a waxy feel. It possesses excellent resistance to most chemicals below 60 °C. Above this temperature the polymer becomes soluble in hydrocarbon and chlorinated hydrocarbon solvents. It exhibits good barrier properties to water vapour but less so for other gases such as oxygen. Its low softening point prevents it being steam sterilised and in the presence of some polar chemicals it is subject to stress cracking. It is readily converted into a lightweight film for pre-packed fresh and frozen foods, and for applications where heat-sealability is required. It is also easily coated onto other materials such as paper and aluminium. LDPE is manufactured into bags and blow-moulded into containers for storage purposes, particularly at low temperatures.

In contrast, high density polythene (HDPE) produced by the low temperature and pressure (50–75 °C, 10 atm) processes using Ziegler catalysts, is stiffer, harder, less transparent, and has a less waxy feel. It has better resistance to oils and greases, a higher softening point but lower impact strength and permeability to water vapour. HDPE is widely used in the production of bottles, bags, tubs, crates, trays, and other household utensils, etc. It has the advantage that it can be steam sterilised.

Polypropylene

Propylene is the next member in the series of olefins and has the structural formula:

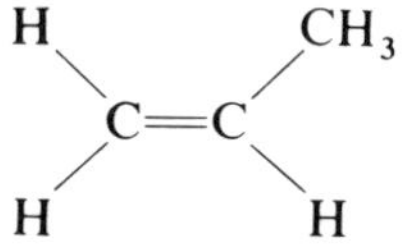

$$\begin{array}{ccccc} H & & & & CH_3 \\ & \diagdown & & \diagup & \\ & & C{=}C & & \\ & \diagup & & \diagdown & \\ H & & & & H \end{array}$$

Generally, polymerisation of such a molecule will produce an irregular macromolecule in which the CH_3— groups are randomly distributed on either side of the main chain as shown below:

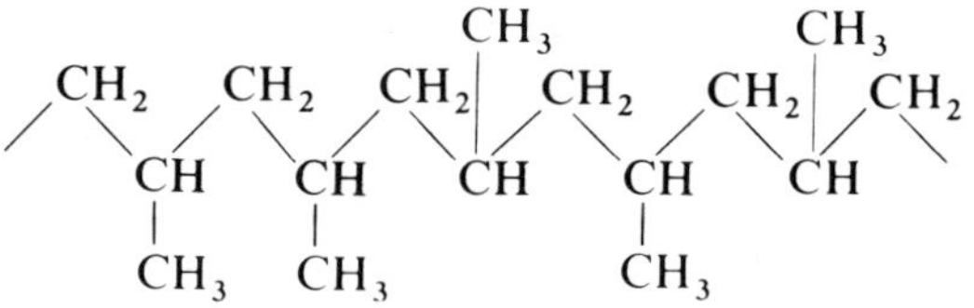

$$\begin{array}{cccccccccccc} & & & & & CH_3 & & & & CH_3 & & \\ & CH_2 & & CH_2 & & CH_2 \,|\, & CH_2 & & CH_2 & | & CH_2 & \\ \diagup & \diagdown & \diagup & \diagdown & \diagup & \diagdown\,|\,\diagup & \diagdown & \diagup & \diagdown & |\,\diagup & \diagdown & \\ & & CH & & CH & & CH & & CH & & CH & \\ & & | & & | & & & & | & & & \\ & & CH_3 & & CH_3 & & & & CH_3 & & & \end{array}$$

Atactic form of polypropylene

This structure is known as the atactic form.[1] However, by using stereospecific catalysts and low pressures it is possible to prepare an isotactic form of polypropylene in which the CH_3— groups are largely ordered onto the same side of the polymer chain, as shown below:

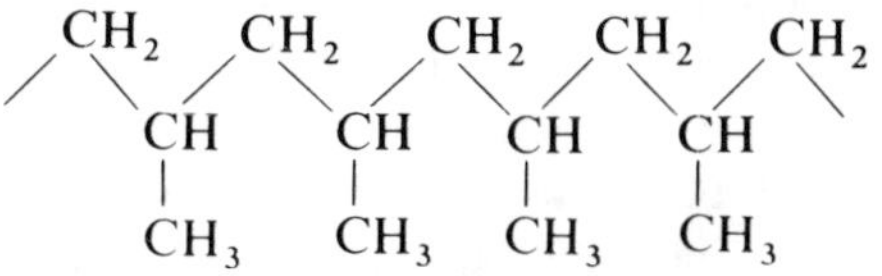

$$\begin{array}{cccccccccc} & CH_2 & & CH_2 & & CH_2 & & CH_2 & & CH_2 \\ \diagup & \diagdown & \diagup & \diagdown & \diagup & \diagdown & \diagup & \diagdown & \diagup & \diagdown \\ & & CH & & CH & & CH & & CH & \\ & & | & & | & & | & & | & \\ & & CH_3 & & CH_3 & & CH_3 & & CH_3 & \end{array}$$

Isotactic form of polypropylene

The regular arrangement of the methyl groups produces a polymer with a higher crystallinity, although not as high as HDPE. In fact, in the isotactic form the carbon atoms are arranged in a helical chain with all the methyl groups projecting on the outside of the helix.

This polymer is harder and has a higher softening point than HDPE and greater resilience but lower shock resistance, especially at low temperatures. It is not subject to environmental stress cracking, possesses good resistance to most chemicals except hot aromatic and chlorinated hydrocarbon solvents and has a permeability intermediate between LDPE and HDPE. Its surface hardness and gloss enables it to be printed upon without difficulty. It is widely used in the manufacture of injection-moulded containers for food storage applications and in extruded form for drinking straws. Further uses arise in the form of biaxially oriented film. The film is stretched in two directions at right angles under suitable temperature conditions. The resulting film has improved gloss and clarity, impact strength and barrier properties to water vapour and oxygen. Its main use in food packaging is as an alternative to regenerated cellulose for the wrapping of snack foods and biscuits.

Polyvinyl Chloride (PVC)

A number of plastics are available in which one hydrogen atom of ethylene has been substituted by a halogen or other group. These constitute the vinyl plastics, as they are prepared from monomers containing the vinyl grouping $CH_2{=}CHR$ (see Fig. 1). Vinyl chloride monomer has the following structural formula:

$$\begin{matrix} H & & & & H \\ & \diagdown & & \diagup & \\ & & C{=}C & & \\ & \diagup & & \diagdown & \\ H & & & & Cl \end{matrix}$$

The polymer exists in the atactic form and, therefore, has a low crystallinity. The substitution of chlorine atoms into the polymer chain increases the polarity of the molecule and the formation of stronger interchain forces. It is a hard, stiff, but clear and glossy material with excellent moisture resistance and low gas permeability. Hence, it is suitable for the packaging of carbonated drinks, mineral waters and cooking oils. Incorporation of plasticisers (usually aromatic esters) softens the resulting film and makes it more flexible but lower in tensile strength, depending on the amount of plasticiser added. This product is widely used in the biaxially oriented form, for the shrink wrapping of meat and cheese. PVC resists attack by most solvents apart from the cyclic ketones, tetrahydrofuran, some chlorinated hydrocarbons and dimethylacetamide. It possesses good resistance to oils and fats and the presence of chlorine makes it less flammable. PVC film or sheet is thermoformed into nests, tubs and trays for a wide range of products, particularly chocolates and cakes.

Vinylidene Chloride (VDC)

This monomer contains an additional chlorine atom substituted into the ethylene molecule:

$$\mathrm{H_2C{=}CCl_2}$$

The polymer is a hard solid with a high degree of crystallinity resulting from its structure and head-to-tail polymerisation pattern. It is insoluble in most solvents and has a very low water absorption. It produces a clear film with outstandingly low permeability to vapours and good strength properties. It is most often used as a copolymer with PVC for the coating of paper, regenerated cellulose and polypropylene where good barrier properties are required. PVDC lowers the working temperature of PVC, facilitating the production of large rigid sheets with good moulding properties.

Polytetrafluoroethylene (PTFE)

In this case the ethylene molecule has been fully substituted with fluorine atoms, as indicated below:

$$\mathrm{F_2C{=}CF_2}$$

The resulting polymer is highly crystalline with a very high molecular weight. It consists of stiff, unbranched chains. F—C and C—C bonds are amongst the strongest encountered in chemistry and this is reflected in the extraordinary chemical inertness of the product. The polymer is smooth and waxy and usually grey in colour. It has a very low coefficient of friction producing a 'non-stick' surface and will withstand a wide range of temperatures. Hence, its principal food contact use is as a coating on frying pans and other cooking ware.

Polystyrene (PS)

Styrene consists of an ethylene molecule in which one of the hydrogens has been substituted by a phenyl radical. This is illustrated below:

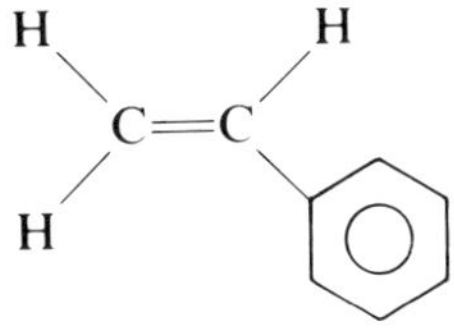

$$\begin{array}{ccc} H & & H \\ & C{=}C & \\ H & & C_6H_5 \end{array}$$

Polymerisation of styrene produces an atactic form and the bulky nature of the phenyl radical prevents close packaging of the macromolecular chains. Hence, it is largely amorphous. Articles manufactured from polystyrene have a characteristic hollow metallic ring when dropped onto a hard surface. The polymer is transparent with a high refractive index and possesses good barrier properties to gases, except for water vapour. It dissolves in higher alcohols, ketones, esters, and aromatic and chlorinated hydrocarbons. Its chief drawback is its brittleness and, hence, it is often copolymerised with butadiene and/or acrylonitrile. Thus it is possible to prepare a wide range of polymers with much improved mechanical properties. The main use of these polymers is as tubs for dairy products, e.g., yoghurt, ice cream, cottage cheese or cream, vending cups and, in the translucent form, for jams. Toughened polystyrene is also used for refrigerator linings. The film is brittle unless biaxially oriented.

NON-ETHYLENIC THERMOPLASTICS

Polyamides (Nylons)

These comprise a series of polymers prepared by the condensation (as opposed to addition polymerisation reactions described previously) of a diacid and a diamine. The first compound produced was nylon 6,6 made from adipic acid and hexamethylene diamine. The reaction can be represented as follows:

$$HN_2(CH_2)_6NH_2 + COOH(CH_2)_4COOH \rightarrow [\text{—}NH(CH_2)_6 . NHCO(CH_2)_4CO\text{—}]_n$$

Other forms of nylon have been prepared and are distinguished by figures denoting the number of carbon atoms in each molecule of amine and acid used for synthesis. Some polymers in this series are obtained by the self-condensation of one molecule only, e.g. nylon 6 prepared from ω-amino caproic acid.

$$NH_2(CH_2)_5COOH \rightarrow [-NH(CH_2)_5CO-]_n$$

The amido group (—NH.CO—) is strongly polar and promotes strong hydrogen bonding between $>$CO and $>$NH groups in adjacent polymer chains. Hence, the nylons are strong, tough, highly crystalline materials with high melting and softening points. Hence, their use in boil-in-the-bag type applications for modern convenience foods. They absorb moisture depending on the relative proportions of hydrocarbons and amido groups in the macromolecule. Characteristic physical properties include low friction, high abrasion resistance and low gas permeability.

Polycarbonates

Polycarbonates are linear polyesters of carbonic acid with aliphatic or aromatic dihydroxy compounds with the general formula:

$$H\left[ORO-\overset{\overset{\displaystyle O}{\|}}{C}\right]_n OROH$$

the basic reaction can be written as:

$$\underset{}{HO-R-OH} + \underset{\text{phosgene}}{2COCl_2} \longrightarrow \underset{\text{bis chloro formate}}{Cl-\overset{\overset{\displaystyle O}{\|}}{C}-ORO-\overset{\overset{\displaystyle O}{\|}}{C}-Cl} + 2HCl$$

$$C-\overset{\overset{\displaystyle O}{\|}}{C}-ORO-\overset{\overset{\displaystyle O}{\|}}{C}-Cl + HO-R-OH \rightarrow H\left[ORO-\overset{\overset{\displaystyle O}{\|}}{C}\right]_n OROH + 2HCl$$

A wide range of aliphatic and aromatic dihydroxy compounds have been tested in this way, to produce polymers spanning a melting range up to 300 °C, but the most important are those obtained from bisphenol A, as shown below.

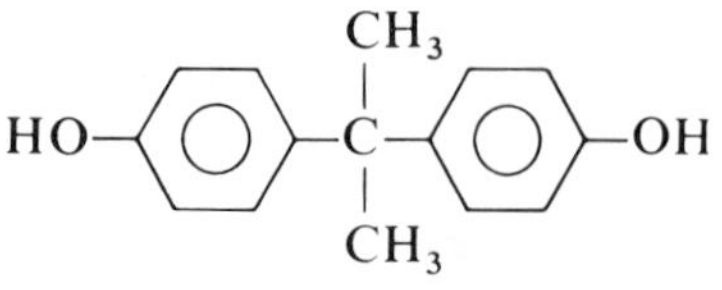

$$HO-C_6H_4-\underset{\underset{\displaystyle CH_3}{|}}{\overset{\overset{\displaystyle CH_3}{|}}{C}}-C_6H_4-OH$$

Most commercial polycarbonates are tough, stiff, hard and transparent. Their physical properties are maintained over a wide temperature range depending on the molecular weight. Polycarbonates are also readily soluble in chlorinated hydrocarbon solvents. In food packaging applications they are used for tableware, fruit juices, beer and for containers for automatic dispensers and for baby feeding bottles.

COPOLYMERS

The monomers discussed above polymerise to produce a single repeating unit and are known as homopolymers. In some instances polymers are prepared either by the addition reaction of more than one monomer or by condensation of three different monomers. In either case, more than one repeating unit will be obtained and the product is known as a copolymer. The properties of a copolymer often differ markedly from either those of the individual homopolymers themselves, or from mixtures or blends of the homopolymers. The properties of copolymers depend not only on the composition of the monomers used, but also on their arrangement along the chain. Along single chains the monomers can be arranged in a regular repeating unit, for example:

—A—B—A—B—A—B—A—B—

or in a completely random manner. Alternatively, 'block' copolymers of the type

—A—A—A—A—B—B—B—B—

are known as well as 'graft' polymers containing side chains built onto reactive centres along the main chain.

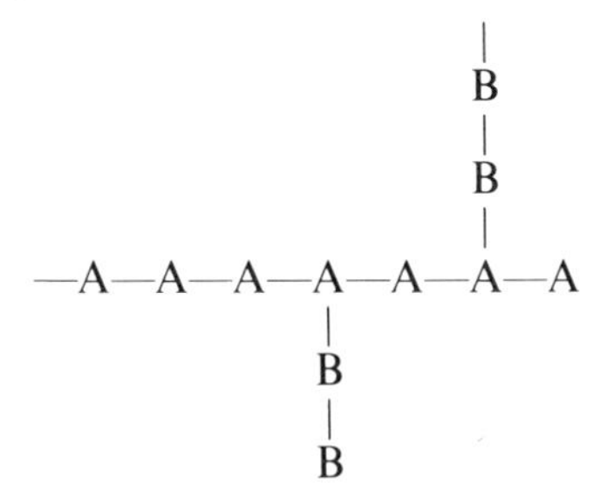

A few examples of copolymers used in the food packaging industry will now be discussed.

Ethylene-Vinyl Acetate (EVA)
Most EVA copolymers contain up to 20 per cent of vinyl acetate and, hence, their properties are very similar to those of LDPE. However, some improvement in transparency and physical properties is obtained, particularly low temperature flexibility, at a cost of higher permeability to water vapour and other gases.

Vinyl Chloride Copolymers
Vinyl chloride is frequently copolymerised with vinyl acetate, vinylidene chloride, polypropylene and acrylonitrile. Vinyl acetate adds bulky groups to the chain, keeping neighbouring chains further apart, thus increasing the flexibility but lowering the softening point. Hence, the copolymer possesses better flow properties whilst remaining tough. VC/VDC copolymers are mainly used as films or coatings where low permeability to water vapour and other gases is required. They are used to improve the barrier properties of materials such as paper, polypropylene and cellulose film. Some VC/PP copolymers containing up to 10 per cent PP are used for making bottles.

Polystyrene Copolymers
Copolymers containing varying amounts of the monomers styrene, acrylonitrile and butadiene have been prepared. Styrene and acrylonitrile (SAN) copolymers are used for some domestic kitchen utensils, rigid containers, bottles and closures. Acrylonitrile butadiene styrene polymer (ABS), with all three monomers present, is used as tubs, trays, boxes, etc., and for liners in refrigerators. The resulting polymers possess much improved high impact properties compared to polystyrene alone. High nitrile resins have excellent gas barrier properties, which make them suitable containers for carbonated beverages such as beer and mineral waters.

LAMINATES

As can be seen from Table 1, no single polymer possesses all the desirable properties that may be required and, as the range of possible applications in the food industry is so diverse, there can be no such thing as the ideal universal packaging material. Optimum properties in each individual case can frequently only be achieved by recourse to several polymers and even non-plastic materials such as aluminium, or paper, in combination. Such combinations, or laminates, consist of layers of individual materials on top of each other and bonded together by adhesives. All-plastic laminates may also be produced as a composite film by co-extrusion or by coating.

OTHER CONSTITUENTS OF PLASTICS

All plastics, in addition to the basic polymer, contain a number of other substances added either deliberately during manufacture and processing or, unavoidably, as residues from polymerisation reactions. Polymers themselves, being of very high molecular weight, inert and of limited solubility in aqueous and fatty systems, are unlikely to be transferred into food to any significant extent. Even if fragments were accidentally swallowed, they would not react with body fluids present in the digestive system. Concern over the safety-in-use of plastics as food packaging materials arises principally from the possible toxicity of other low molecular weight constituents that may be present in the product and, hence, leached into the foodstuff during storage. As stated above, such constituents arise from two sources:

(1) Polymerisation residues including monomers, oligomers (with a molecular weight up to about 200), catalysts (mainly metallic salts and organic peroxides), solvents, emulsifiers and wetting agents, raw material impurities, plant contaminants, inhibitors, decomposition and side reaction products.

The more volatile gaseous monomers, e.g. ethylene, propylene, vinyl chloride, usually fall in concentration with time, but very low levels may persist in the finished product almost indefinitely. Styrene and acrylonitrile residues are more difficult to remove.

(2) Processing aids such as antioxidants, antiblock agents, antistatic agents, heat and light stabilisers, plasticisers, lubricants and slip agents, pigments, fillers, mould release agents and fungicides.

Such constituents may be added to assist production processes or to enhance the properties and stability of the final product. They may be present in amounts varying from only a few parts per million up to several per cent.

Since compounds of the first group are present inadvertently there is not a lot that can be done to remove them. However, the efforts made by industry to reduce viny! chloride monomer levels, in particular, illustrates the advantages of optimum manufacturing processes on the purity of the final product. Chemicals added deliberately during formulation to alter the processing, mechanical or other properties of the polymer are likely to be present in greater amounts than polymerisation residues and should be subject to strict quality control. They are normally restricted to compounds appearing on an approved list for food contact use. Whilst the greatest danger to human health from such additives would be expected to arise at the factory where large quantities have to be manipulated by plant operatives, extraction and migration tests on the finished product are now considered to be essential by most legislative authorities for the protection of the consumer. The large number of individual compounds likely to be encountered in use, presents many difficulties for the analyst. Analytical methods for the determination of a number of individual additives in extractants, in food simulants and in polymers have been reviewed by Crompton.[2] Methods for the identification and analysis of polymers can be found in the work of Haslam[3] and Van Gieson.[4]

A brief mention of the function of some major additives is now presented.

Antiblock Agents

These are added to roughen the surface of thin films and, hence, prevent them sticking together during machine processing. Silica is most commonly used because its poor solubility in most polymers helps to increase the surface concentration and so introduce irregularity. Similarly, *slip* additives such as fatty acid amides, are used to reduce mobile frictional forces.

Antioxidants

These prevent degradation of the polymer by reacting with atmospheric oxygen during moulding operations at high temperatures or when used in contact with hot foods; and to prevent embrittlement during storage. Derivatives of phenols and organic sulphides are most frequently used. Some of these compounds are classified as *heat stabilisers*.

Antistatic Agents

Since all plastics are good electrical insulators (and are in fact used on a large scale for this purpose) they will retain electrostatic charges produced by friction from contact with processing machinery. Accumulation of static electricity can cause problems through the pick-up of dust, adhesion between layers or particles of plastic, sparking, electrical shock and possibly fire hazards. Most antistatic agents are glycol derivatives or quaternary ammonium compounds, which both increase the electrical conductivity and plate-out onto the surface of the plastic.

Lubricants

These are added to reduce frictional forces and are usually low to medium molecular weight hydrocarbons. They should possess good solubility in the plastic, low volatility and be relatively stable compounds.

Plasticisers

These are added to make the product more flexible and less brittle. They are usually high molecular weight esters, e.g. phthalates.

UV Stabilisers

These may be needed to protect the product from deterioration by sunlight, or even supermarket lighting. Products containing vitamin C are particularly susceptible to this form of deterioration.

REFERENCES

1. NATTA, G., Isotactic polymers, *Chem. Ind.*, London, 1957, **957,** 1520.
2. CROMPTON, T. R., *Chemical analysis of additives in plastics*, Second edn, Pergamon, Oxford, 1978.
3. HASLAM, J. and WILLIS, H. A., *Identification and analysis of plastics*, Iliffe, London, 1965.
4. VAN GIESON, P., Here's a quick, easy way to identify films, *Package Eng.*, 1969, **4,** 76.

SUGGESTIONS FOR FURTHER READING

BILLONEYER, F. W., *Textbook of polymer science*, Interscience, London, 1962.

BRISTON, J. H., *Plastics films*, Iliffe, London, 1974.

BRISTON, J. H. and KATAN, L. L., *Plastics in contact with food*, Food Trade Press, London, 1974.

COUZENS, E. G. and YARSLEY, V. E., *Plastics in the service of man*, Penguin Books Ltd, Harmondsworth, 1956.

Encyclopedia of polymer science and technology, Interscience, New York, 1964.

FLORY, P. J., *Principles of polymer chemistry*, Cornell University Press, New York, 1953.

IARC, *Some monomers, plastics and synthetic elastomers*, Monograph Vol. **19,** International Agency for Research on Cancer, Lyon, 1979.

MOORE, W. R., *An introduction to polymer chemistry*, University of London Press, London, 1963.

PINNER, S. H. (Ed.), *Modern packaging films*, Butterworths, London, 1967.

SAUNDERS, K. J., *Organic polymer chemistry*, Chapman and Hall, London, 1973.

Chapter 3

DETERMINATION OF MONOMERS

INTRODUCTION

Increasing concern over the possible effects of vinyl chloride monomer (VCM), and some other monomers too, on human health (see Chapter 4) has led to the promulgation of regulations in several countries. These regulations protect both workers in and around production and processing plants, as well as the resident population in general, from exposure to residual levels of monomers arising from contact with materials and articles used in the food industry. Compliance with these limits presupposes the existence of simple, sensitive and reliable analytical methods for the control of plant atmospheres, effluents, plastic articles and residue levels migrating into foodstuffs (Chapter 7). Despite the fact that all monomers are under suspicion toxicologically, most of the analytical studies in the literature relate to the determination of vinyl chloride. This work will now be reviewed critically.

ENVIRONMENTAL AND PLANT MEASUREMENTS

In order to check that permitted occupational exposure levels have not been exceeded, analytical methods are required that can be used for a large number of 'snap' samples, often taken at locations some distance away from the laboratory. Alternatively and additionally, continuous monitoring of selective sites around the production plant may also be required. Finally, portable devices suitable for wearing by individual workers have been developed to enable an assessment of personal exposure to be made.

Sampling Methods Available

The sampling of atmospheres in and around the production plant creates its own set of problems in addition to those resulting from the subsequent analytical determination step. 'Snap', or 'grab', samples taken using evacuated containers give an indication of the VCM concentration at one particular place and point in time. Delays in transit to the laboratory, and in analysis at the laboratory may mean that a long time elapses before the result is known and before any corrective action can be initiated at plant level. Hence, this method contributes very little to the efficient running of the plant or to personnel protection. Furthermore, the large number of samples that would have to be taken to provide an effective monitoring programme would impose a heavy load on expensive analytical facilities, including skilled technical staff.

Alternatively, time-averaging techniques can be used. This type of screening involves sampling tubes, traps (usually containing charcoal) or inflatable bags, in conjunction with a portable battery-driven pump which abstracts a continuous stream of the atmosphere over a period of time which may be as long as 8 or even 24 h. Portable devices that can be carried on the operative's body throughout a working shift have also been developed. This technique thus provides information of monomer levels over a much greater period of time, with the minimum of analytical effort, by comparison with instantaneous 'snap' sampling methods. However, it does have a very great limitation in that the results obtained are an *average* of VCM concentrations throughout the sampling period. Thus, VCM levels above the legal limits, or in the potentially hazardous range, even though for only short periods of time would not be revealed by this technique.

Undoubtedly, continuous monitoring techniques which provide a permanent record of both average values as well as some indication of fluctuations over the period of a shift, and from day-to-day, constitute the most satisfactory means of screening production plants and their associated environment. Atmospheric levels of VCM are known to fluctuate wildly and, hence, the result obtained will vary with the time of sampling. Figure 1 illustrates a comparison of data obtained from both continuous monitoring and time-average sampling equipment over a period of 2 h. Whilst the average over the 2 h period was only 1 ppm, the figure shows that, for short periods, peak values greater than 5 ppm were experienced as well as extended exposure periods above the 1 ppm level. Continuous monitoring techniques also give virtually instant read-outs of VCM levels and, hence, enable the fastest possible corrective action to be

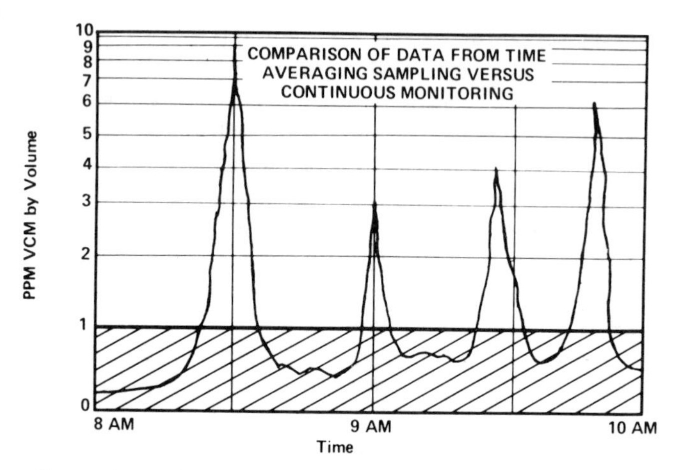

FIG. 1. Comparison of data from time-averaging sampling versus continuous monitoring. Reprinted with permission from *American Laboratory*, vol. 7 no. 12, 1975, Copyright 1975 by International Scientific Communications, Inc.

taken when leaks occur on the plant site. Alarm signals for the protection of personnel can also be built into these detection systems.

Analytical Techniques

Whilst a wide variety of colorimetric, electrometric and spectrophotometric techniques have been proposed for the detection of vinyl chloride monomer,[1,2] gas chromatography has proved to be the most widely used. VCM (with a b.pt. of $-13{\cdot}5\,°C$) presents no real problem regarding its separation from the constituents of the industrial atmosphere, although impurities in some solvents commonly used for the analysis of VCM, may require that normal length analytical columns must still be employed. Guidelines as to suitable stationary phases and operating conditions can be found by reference to the bibliography. These will, however, depend to some extent on the other constituents present in the sample. The flame ionisation detector (FID) is most commonly used for vinyl chloride analysis and is sufficiently sensitive for measurements down to 0·1 ppm to be made. However, since it responds to all organic compounds it possesses no discriminatory power, particularly for industrial atmospheres which may contain hydrocarbons, acetaldehyde, other chlorinated compounds, etc., as well as VCM. The electron-capture detector (ECD), whilst more specific for VCM, lacks sensitivity since the molecule contains only a single chlorine atom. More sophisticated detection systems based on mass spectrometry,[3] chemiluminescence[4] or carbon monoxide/carbon dioxide lasers[5] have also been described. Some of these techniques are discussed more fully in the section on analysis of polymers and foodstuffs.

Infra-red spectrometry provides an alternative measuring technique, sensitive down to about 1 ppm. It has the advantage of being more specific for VCM although its lack of sensitivity requires that long (up to 20 m) multi-path cells are used. VCM exhibits strong absorption bands at 5·15, 9·8, 10·9 and 13·9 μm but the bands (at 10·9 μm) due to the out of plane vibrations of hydrogen atoms in the vinyl group are probably the most specific. A selective radiation type of instrument without monochromator can be used for continuous monitoring but many substances can interfere with the analysis.

A continuous monitoring system dependent on the photo-electric detection of light energy reflected from an impregnated paper tape has been developed by Universal Environmental Instruments.[6] This instrument produces a response, specific to VCM (apart from other unsaturated chlorinated hydrocarbons such as trichlorethylene and vinylidene chloride) and can detect levels down to 0·1 ppm. Other advantages include ease of operation and calibration, with a direct read-out in ppm of VCM. The sampling system incorporates a soda lime scrubber to remove acidic gases and free chlorine, followed by an oxidant to convert VCM into free chlorine which then reacts with the chemically sensitised paper tape.

Adsorption on Charcoal

Many methods for the determination of VCM in the atmosphere are based on the adsorption of the compound in a metered stream of air passing through tubes containing activated charcoal, followed by solvent extraction or desorption prior to analysis by gas chromatography. Hill *et al.*[7] examined 20 different solid sorbents of which coconut shell charcoal was found to be the most practical. As expected, the breakthrough volume increased inversely with the concentration of VCM and the sample flow rate. Desorption using carbon disulphide gave recoveries usually better than 80 per cent, with a relative standard deviation of 7·5 per cent. The vinyl chloride adsorbed on charcoal was stable for three weeks when stored at ambient or sub-ambient temperatures. Cuddeback *et al.*[8] had earlier obtained similar results. They found that 100 mg of activated charcoal has a capacity of 65 μg of VCM with a two-fold safety factor when sampling air containing VCM at the 5 ppm level. Ninety per cent recoveries were achieved after storage for two weeks. These studies have shown that the preparation and the previous history of the charcoal are important. Batch to batch variations have been reported by several workers and the relative humidity of the atmosphere[9] as well as packing characteristics, may also be important. Miller *et al.*[10] have recently described the preparation of the adsorption tubes and the pretreatment of charcoal. Over a 24 h sampling period, a limit of detection as low as 0·005 ppm v/v has been achieved. Ahlstrom *et al.*[11] have examined some other solid adsorbents. Carbosieve B was found to be too strong an adsorbent whilst Porapak Q, Chromosorb 102 and Tenax did not absorb VCM in measurable amounts from the atmosphere under the conditions chosen. On the other hand, Tenax GC (a porous polymer of 2,6-diphenyl-*p*-phenylene oxide) has been used successfully by a number of other workers for trapping a wide range of organic compounds in the atmosphere. The properties of Tenax as an adsorbent have been discussed by Bertsch *et al.*;[12] and Ives[13] was able to trap levels of VCM as low as 60 and even 6 ppb (v/v), recovering 90–100 per cent. He found that it was essential to cool the Tenax with dry ice, before flushing the entire sample onto the gas chromatographic column by heating.

Further laboratory and field investigations of a number of experimental parameters that might influence adsorption onto charcoal have been reported by Miller *et al.*[10] Interference by a range of possible atmospheric contaminants was established as negligible. Carbon disulphide has found almost universal acceptance as the solvent for desorption, although dichloromethane has been suggested as a possible alternative.[10] Solvent temperature and volume have an effect on desorption efficiency. Addition of charcoal to solvent was preferred by Hill *et al.*[7] and 10 min was sufficient time for weights of VCM up to 2 μg to be desorbed. Since mixing of charcoal with carbon disulphide is an exothermic process, some workers recommend cooling to 0 °C to reduce losses of VCM.

The principal disadvantages of the charcoal adsorption and solvent desorption system are:

(1) the dilution factor introduced by the use of solvent for desorption;
(2) the impurities present in carbon disulphide together with its high toxicity (TLV = 20 ppm), flammability and smell;
(3) losses of VCM into the headspace above the solvent may occur.[8]

Low boiling impurities occurring in carbon disulphide may interfere with the VCM peak during determination by GLC. High boiling impurities may also be present and will leave the chromatographic column only slowly, thus lengthening the analysis time unless backflushing facilities are available or the oven temperature is raised at the end of the analysis. The charcoal adsorption tubes are normally arranged in two sections, as a check on breakthrough during sampling of atmospheres containing high levels of VCM. Commercial tubes normally contain 150 mg of adsorbent (20/40 mesh) packed into a glass tube (i.d. = 4 mm) in two sections, separated by a polyurethane plug; 100 mg in the front section and 50 mg at the back as a check on the breakthrough of VCM. However, unless the tubes are refrigerated at −20 °C between sampling and analysis, migration of VCM from the front to the back section of the adsorption tube may occur. Hence, tubes stored at ambient temperatures must be analysed as a combined unit. Normally 5 litres of atmosphere is sampled at a rate of 0·05 litre/min and levels of VCM as high as 200 ppm can be tolerated without significant breakthrough as long as the tubes are packed without channelling.

VCM Standards

Caution

VCM is a gas at ambient temperatures and a known carcinogen by the inhalation route. Consequently, operations with VCM must be carried out in a well ventilated fume cupboard and operators must wear gloves made of a material, such as neoprene, that does not absorb vinyl chloride readily.

In order to monitor the accuracy and reliability of the laboratory analytical technique, it is necessary to prepare standard reference atmospheres containing known quantities of VCM. It is not easy to prepare standard atmospheres that are stable and reproducible at the very low concentration levels required for calibration purposes, for a gas such as VCM. Dynamic generation systems have been described by Ash and Lynch,[14] and by Cuddeback *et al.*[8] Standard samples of air containing declared levels of VCM are available commercially (e.g. Air Products Limited).

The problems in preparing standards of VCM have been discussed by Ravey and Klopstock.[15] They produced mixtures of VCM and nitrogen in the range 1 to 100 ppm of VCM, in glass hypovials, sealed with neoprene septa and aluminium crimp caps. After five days the 100 ppm standard had fallen to 97 ppm, but at levels of 1 and 10 ppm, only 80–90 per cent of the initial concentration remained after 24 h. Standards of VCM may also be obtained in solution and this may improve the chromatography, since smaller injection volumes are required. Ravey and Klopstock[15] compared 1,2-dichloroethane with tetrahydrofuran (THF). The former solvent produces more stable standard solutions but has to be warmed to 40 °C to dissolve PVC, which is soluble in THF at ambient temperatures. Internally silvered glassware for use with THF solutions of VCM was recommended by these authors since, it was claimed, adsorption of the VCM onto the walls of the container occurred at low concentrations.

A collaborative study of vinyl chloride charcoal tubes prepared by a permeation gas generation system for reference purposes, has been published by Shou-Yien Ho.[16] At concentrations of 12 and 20 μg VCM per tube, very small generation and replication errors were found. However, the variation between laboratories was as high as 15 to 20 per cent, probably resulting from errors in the preparation of the standard calibration curve. Further, comments on the preparation and use of standard VCM solutions can be found in the section (below) dealing with the analysis of polymers and foodstuffs. A comprehensive review of the problems encountered in the measurement of VCM in the atmosphere has been published by Lande.[17]

Personal Monitoring Systems

These sampling techniques consist of a small battery-driven pump which pulls air from the working zone through a metering device and a collection tube usually filled with charcoal. Subsequent analysis of the adsorbed VCM on the charcoal is carried out at the laboratory as described above. An alternative system based on permeation, or diffusion, has also been developed.[18] Vinyl chloride in the atmosphere dissolves in a silicone membrane exposed to the atmosphere. After passing through the membrane, VCM is held, by adsorption, on activated charcoal. Trials have established that factors such as air movement, humidity, the presence of other contaminants or even temperature over the range 0 to 40 °C, have little significant effect on the results obtained. Calibration is effected by exposure to standard atmospheres containing known amounts of VCM for a given time, whence:

$$K = ct/w$$

where, K = permeation constant, c = concentration of vinyl chloride in the atmosphere, t = time of exposure and w = weight of VCM adsorbed and found by analysis.

A small positive bias by the permeation monitor was detected in field studies, but the system is cheaper, less subject to failure, easier to wear and generally more convenient in use than most alternative methods of personal monitoring.

Effluent Samples

The polymerisation of VCM to PVC is carried out in autoclaves in which the monomer is dispersed in the form of tiny droplets in an aqueous medium. At the end of the reaction, the slurry is centrifuged and the autoclave is cleansed with high-pressure water jets. Hence, a high volume of water is used and it is important to check the VCM content of any effluents discharged. The earlier usage of VCM as a propellant in aerosol cans may possibly constitute a secondary source of pollution of underground water systems, where solid domestic waste has been tipped. Whilst PVC is used in some domestic supply systems, there is little likelihood of significant transfer, since the solubility of VCM in water is so low.

Methods for plant effluents have been described by Thain,[2] based on pre-concentration followed by GLC determination using the headspace technique (see later). Quantitative methods for low levels of VCM in water have also been described by Bellar *et al.*[19] and by Dressman and McFarren.[20] Direct injection methods using gas chromatography with FID or halogen-specific detectors are not sufficiently sensitive without pre-concentration of the aqueous sample. Solvent extraction may not be satisfactory since its efficiency will be relatively low, and other co-extractants may interfere in the gas chromatographic stage. The method developed by Bellar *et al.*[19] involves purging with nitrogen, followed by trapping on silica gel or Carbosieve B and subsequently desorption at 150 °C. The method was found to be satisfactory within the range 4–40 μg/litre for VCM in tap, river or sea water. Halogen-specific detectors were used and GC/MS was available for confirmation. Even lower detection limits were achieved by Dressman and McFarren[20] using a slightly modified sampling technique and a microcoulometric titration detector; 0·1 μg/litre could be detected in a 5 ml sample. It is important for all water samples to be taken from full bottles containing no airspace above the water level, since VCM is readily lost to the atmosphere above aqueous solutions.

TRACE LEVELS IN POLYMERS AND IN FOODS

Many similar problems to those already described for atmospheres and effluents are encountered in the determination of VCM in polymers or in food. The major difference arises from the technique used to separate the VCM from the sample, either polymer or foodstuff. The major analytical challenge results from the low levels of VCM encountered, viz. 1 ppm or less in the case of finished polymers and 10 ppb or less in foods. Polymers have to be dissolved (or dispersed) in order to release any free monomer prior to analysis. Foodstuffs are even more complex (and variable) in chemical

TABLE 1
PROPERTIES OF SOME MONOMERS

	Ethylene	Propylene	Butadiene	Styrene	Vinyl chloride	Vinylidene chloride	Acrylonitrile	Vinyl acetate
Structural formula	$CH_2{=}CH_2$	$CH_3CH{=}CH_2$	$CH_2{=}CH{-}CH{=}CH_2$	$PhCH{=}CH_2$	$CH_2{=}CHCl$	$CH_2{=}CCl_2$	$CH_2{=}CHCN$	$CH_2{=}CHOOCCH_3$
Molecular weight	28·05	42·06	54·09	104·14	62·5	96·95	53·03	86·09
Boiling point, °C (760 mm Hg)	−103	−47	−4·4	145	−13·9	31·5	77·5	72·7
Freezing point, °C	−169	−185	−109	−32	−152	−123	−83	−100
Specific gravity	1·21	1·5	—	—	2·15	3·4	1·2	2·97
Density, g/ml (20 °C)	0·30	0·52	0·62	0·91	0·98	1·22	0·81	0·93
Refractive index, n_D^{20} (20 °C)	1·363	1·356 7 (−70 °C)[a]	1·429 3 (−25 °C)[a]	1·546 3	1·398	1·424 7	1·391 1	1·395 3
Solubility in water at 25 °C, wt per cent	1·1	44·6	0·11	0·032	0·11	0·021	7·3	2·3
Solubility of water in monomer, per cent	—	—	—	0·07	—	0·035	3·4	0·9
Vapour pressure (mm Hg) 20 °C	38 000	7 600	—	10	2 530	495	80	92
40 °C						720	200	
60 °C							440	
80 °C							815	
Threshold limit value	—	—	1 000 ppm	100 ppm	—	—	20 ppm	0·2 mg/cm³

[a] Values are at temperatures given in parentheses.

composition than polymers and this adds a new dimension to the analytical problem. For this reason, migration experiments to record the transfer of monomer from the plastic into the food are often carried out using simulants in place of real foods, to simplify the analytical determination stage (Chapter 7). However, for surveys of the national diet, to determine intake levels of monomer, direct analysis of the foodstuff has to be employed and the analytical difficulties overcome.

The high volatility of VCM is a considerable aid to its detection and measurement in polymers and foods by gas chromatography. The development of headspace gas chromatography as a general analytical technique for the determination of volatile constituents of liquids or solids has contributed an important advance and impetus to solving some of the problems referred to above. Whilst other monomers such as vinylidene chloride, acrylonitrile and styrene are much less volatile in comparison with VCM (see Table 1) they too are amenable to determination by headspace techniques, and recent developments will be described later.

Headspace Gas Chromatography (HSGC)

HSGC is a specific form of gas chromatography in which the vapour space above the sample (liquid or solid) enclosed in a gas-tight container (or vial) is taken for analysis; rather than an injection of the sample dissolved in a liquid, as in conventional gas–liquid chromatography (see Fig. 2). The sample may be in the form of a solid, a liquid, or a solid dissolved in a liquid. The vial is maintained in a thermostatted environment until equilibrium of

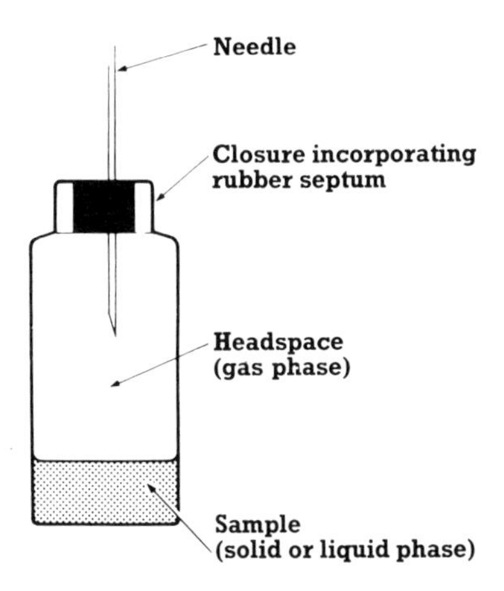

FIG. 2. Sampling vial for HSGC.

the volatile components of the sample, between the gaseous and liquid phases, has been established. Then, a portion of the gaseous phase is transferred to the GLC column. This transfer can be performed manually with a gas-tight syringe under controlled conditions, or automatically by positive pressure valve switching as in the Perkin–Elmer F40, F42 and F45 instruments.[21] Gas-tight syringes have to be used with a great deal of care, particularly in quantitative work. The syringe should be pre-warmed to equilibration temperature, otherwise the equilibrium may be disturbed and condensation can occur. Withdrawing a sample creates a partial vacuum, again disturbing the equilibrium. Equally for trace analysis, adsorption on the walls of the syringe and memory effects can be a problem.

In manual sampling the use of an internal standard is, therefore, essential to correct for variations in the volume of gas injected onto the column. The internal standard selected should be chemically similar to the determinant but clearly separated from it on the chromatogram. It is also necessary to add the same internal standard to all standard solutions used for calibration purposes. Quantitative measurements are then made by computation of peak area (or height) ratios of determinant to internal standard for the test solution and multiplying by the calibration factor, determined from similar ratios for standard solutions containing known amounts of determinant. The concentration of the internal standard used must be such as to produce a peak of similar size and shape to that of the determinant on the same chromatogram.

The electropneumatic dosing system described by Kolb[21] obviates many of these difficulties and has been widely used for forensic blood alcohol analyses and many other applications besides the determination of monomers. The technique has the advantage of being more reproducible and more suitable when large numbers of samples have to be analysed as in quality control applications. Up to 30 samples can be accommodated on the turntable and peak heights can be reproduced with a relative standard deviation of only 0·5 per cent,[21] although the absolute volume of sample injected is unknown. Furthermore, the use of an internal standard is not required.

The chief advantages of the headspace technique compared with conventional gas chromatographic methods are that preliminary procedures, such as extraction, distillation, digestion, etc., to concentrate the trace constituent and remove the sample matrix are not required. Additionally, there is less contamination of the injection port and chromatographic column with macro components, solvents or co-extractants, particularly non-volatile materials, hence increasing the separating power and shortening each chromatographic run. Where liquid solvents are not required for extraction, the use of a gas eliminates the solvent peak (often tailing) from the chromatogram as well as peaks from impurities, since gases are usually available in purer form than liquid solvents. One disadvantage of the technique is that the quantitative interpretation of the results is less certain and the sensitivity cannot be increased simply by taking a larger sample. The technique records the concentration of the trace constituent in the headspace (as its name implies), and, therefore, the relation of this value to the concentration of the same trace constituent in solution, at equilibrium, must be investigated in some detail.

Theoretical Aspects of HSGC

The resulting chromatogram of a headspace analysis represents the composition of the gas phase. The area of each peak produced is proportional to the partial vapour pressure of the compound in the gas phase. As an approximation, the peak height can be used in place of peak area. Hence, the basic quantitative relationships of headspace analysis can be represented as follows:

$$A_i' = P_i' C \qquad \text{Gas phase} \tag{1}$$

where A_i' = peak area (or height) of component i, P_i' = partial vapour pressure above the solution (or sample), and C = the calibration factor.

The partial vapour pressure, in turn, is determined by the concentration of component i in solution or in the sample, and can be found using Henry's law. This states that the mass of gas dissolved, by a given volume of solvent, at constant temperature (T), is proportional to the pressure of the gas in equilibrium with the solution.

The partial vapour pressure (P_i') depends on the mole fraction (x_i) of the solute (i) and the vapour pressure of the pure compound (i) and can be expressed in the form:

$$P_i' = P_i^0 x_i \gamma_i \tag{2}$$

where, x_i = mole fraction of i dissolved in the sample, P_i^0 = vapour pressure of pure component i at temperature (T) and γ_i = the activity coefficient of i and corrects for any deviation from ideal behaviour.

In the case of an ideal solution, $\gamma_i = 1$ and eqn (2) reduces to Raoult's law:

$$P_i = x_i P_i^0$$

However, most solutions exhibit positive or negative deviation from Raoult's law as a result of intermolecular forces. In any case, for trace analysis the solutions are dilute and even though the activity coefficient may vary from unity, the value is usually constant within the practical range of measurement. The activity coefficient is dependent not only on the chemical nature of the component, and other components in the mixture, but also on the temperature, pressure and concentration. Hence, calibration must be carried out with care.

Combination of eqns (1) and (2) produces the basic working equation for all headspace analysis:

$$x_i = \frac{A_i}{C\gamma_i P_i^0} \tag{3}$$

or

$$x_i = \frac{A_i}{C_2} \tag{4}$$

assuming γ_i is constant within the range of measurement. The calibration constant (C_2) is determined by measurements using known amounts of the pure compound (i), in the same sample matrix under identical experimental conditions. This ensures that the activity coefficient is not disturbed from its value (under test conditions) by variations in ionic strength or other changes in the composition of the matrix. Where the composition of the sample is unknown or is unavailable and so cannot be reproduced, the method of additions can be used to compensate for any possible matrix effects. Known additions to further portions of the same sample are made. The sample and enriched samples are then analysed under identical conditions. From the increased peak areas obtained for a given level of enrichment it is possible to calculate the original concentration in the sample. The application of the method of additions to the determination of hydrophilic solutes and benzene in aqueous systems by HSGC has been described by Drozd and Novák.[22,23] Many potential applications of headspace gas chromatography have been discussed by Hachenberg and Schmidt.[24]

Practical Application of HSGC to the Determination of Monomers

This section is restricted to an examination of experimental parameters that are important as a result of the adoption of the headspace method of sampling for analysis. It is assumed that normal gas chromatographic equipment is available and functioning properly, having been adjusted in accordance with the manufacturer's recommendations to produce peaks approximately Gaussian in profile. In this type of work the selection of particular chromatographic columns, with specified stationary phases is not considered to be critical, provided that adequate resolution of the monomer peak (and internal standard peak, if used) from impurities in the solvents or from other co-extractants present in the sample can be achieved. For precise quantitative measurements, baseline separation of the monomer peak from all other peaks must be obtained by the selection of appropriate stationary phases or column conditions. For qualitative work, the use of more than one column containing stationary phases which differ in polarity, is recommended by many workers for confirmation of the identity of suspect monomer peaks. Generally speaking this approach is only satisfactory where significant changes in peak retention times are observed, often involving a reversal of the order of peak elution. Otherwise, only limited additional information is obtained and recourse should be made to more selective detection systems for confirmation of peak identity.

Detectors

The following detectors are widely used in conjunction with headspace sampling for the determination of trace amounts of monomers both in plastics, foods or food simulating solvents:

Flame ionisation detector (FID)
Nitrogen-selective detector (ND)
Electron-capture detector (ECD)
Hall detector
Coulometric detector
Chemiluminescence detector
Mass spectrometer (MS)

The FID is a universal detector for organic compounds.[25] Whilst it is very sensitive and possesses a very wide linear range, it exhibits little specificity. Hence, a number of attempts have been made to use the alternative selective detectors indicated in the list above. The ND is a modified form of the FID often referred to as the alkali flame ionisation detector (AFID), since it comprises a normal FID with a bead of an alkali metal salt, e.g. $SrSO_4$ situated in the flame. This modifies the ionisation processes occurring in the flame, so that response to nitrogen compounds is considerably enhanced by comparison with its response to carbon compounds. The phenomenon was first reported by Giuffrida[26] for phosphorus and work on the detector in the nitrogen mode has been reported by Aue *et al.*[27] and by Brazhnikov *et al.*[28] An alternative type of nitrogen-selective detector was described by Coulson.[29] This operates in two different modes, being selective towards either nitrogen or to the halogens. The eluate from a gas chromatograph is passed through a furnace, where compounds containing nitrogen or halogens are converted catalytically to ammonia or to halogen acids, respectively. The concentration of these species is then monitored using a specially designed conductivity cell. A modified form of this detection system has been described by Hall[30] and used by Ernst and Van Lierop for the determination of VCM[31] and VDC.[32] This detector is much more selective than the FID and is easier to operate in the nitrogen mode than the modified flame detectors.

An alternative detector for halogen-containing compounds was developed by Lovelock.[33] The ECD has been widely used for compounds such as organochlorine pesticides which possess a strong affinity for electrons, but in the case of VCM containing only one chlorine atom per molecule, the sensitivity is no better than the FID. However, other workers[34,35] have used the ECD after chemical conversion of VCM to 1,2 dibromo-1-chloroethane. Williams[34] found that the reaction only went to between 40 to 55 per cent of the theoretical yield but the detector could still be used down to 20 pg (20×10^{-12} g).

Chemiluminescence detectors depend on the reaction between ozone and an olefin, which gives rise to a continuum of radiation with a maximum intensity at 400 nm which can be monitored by a simple light detection system. The technique has been used by McClenny *et al.*[4] for the detection of VCM and it should be possible to adapt it for other monomers such as VDC and AN. In preliminary trials of a pilot system in these laboratories, insufficient sensitivity was achieved to permit analyses at the parts per billion level.

Mass spectrometry is used in this work as a highly specific GLC detector. In most cases single ion monitoring is practised, although where interferences are present it is usually possible to monitor more than one ion from the fragmentation pattern. For compounds which contain chlorine, the naturally occurring isotopic ratio of the Cl^{35} to Cl^{37} atoms is reflected in the occurrence in the fragmentation pattern of ions with parent masses (m/e) of 62 and 64 in the ratio 3:1, thus providing additional confirmation of the true identity of the peak. Where doubt remains, the whole spectrum should be scanned and examined with due regard to the resolution parameters of the instrument in use.

Sensitivity

Whilst the sensitivity and the minimum detectable quantity is a function primarily of the detector, headspace sampling imposes further limitations not encountered in other forms of GLC. For example, increasing the absolute amount of sample in the headspace sampling vial (Fig. 2) does not itself improve the limit of detection, since it is the concentration of monomer in the gas phase and not the total amount in the vial that is being measured. Hence, an increase in sensitivity can only be achieved by changing the value of the partition coefficient, the activity coefficient or by increasing the temperature and thereby increasing the partial vapour pressure of the monomer as indicated by the Clausius–Clapeyron equation:

$$\frac{\mathrm{d}\ln p}{\mathrm{d}T} = \frac{L}{RT^2}$$

where, L = latent heat of evaporation.

An increase in the temperature also shortens the equilibration time required. Steichen[36] showed that some increase in sensitivity for VCM could be achieved by the addition of water, thus decreasing the solubility of the monomer in the solvent by a factor of around two or three. Greater increases in sensitivity were obtained for styrene and butadiene.

Vinyl Chloride Monomer (VCM)

The problem of calibration has already been mentioned in the theoretical section. This process is made more difficult for VCM by the fact that standard solutions are not easy to prepare in the laboratory. Some of these difficulties will now be discussed. Many of the early workers used aqueous solutions of VCM as standards. Since VCM is a gas with a b.pt. of $-13\,°C$ and is only slightly soluble in water, standard solutions tend to be unstable, in use and on storage, unless strict precautions are taken. Bubbling the pure gas through water at room temperature produces a saturated solution containing about 0·2 per cent (w/v) of VCM. However, VCM is readily lost from such a solution by normal laboratory handling operations, e.g. dilution, pipetting or transferring from one vessel to another. Furthermore, in a partially filled container the VCM will be distributed between the solution and the headspace, depending on the temperature.[37] This effect is illustrated in Fig. 3, which shows the concentration of VCM in the headspace at different headspace volumes. Hence, standard solutions of VCM in water should be stored in gas-tight containers filled to leave as little headspace as possible. The containers should be stored in a refrigerator and under these conditions the solutions may remain stable for a month or two.

Dilutions, where necessary, should be made using distilled water that has been chilled previously by storage in a refrigerator. Solutions should be poured carefully down the side of the container with the minimum of agitation. Meticulous attention to practical detail is required to obtain repeatable results. However, as Fig. 3 shows, the partition of VCM into the headspace is more favourable in aqueous systems compared to those

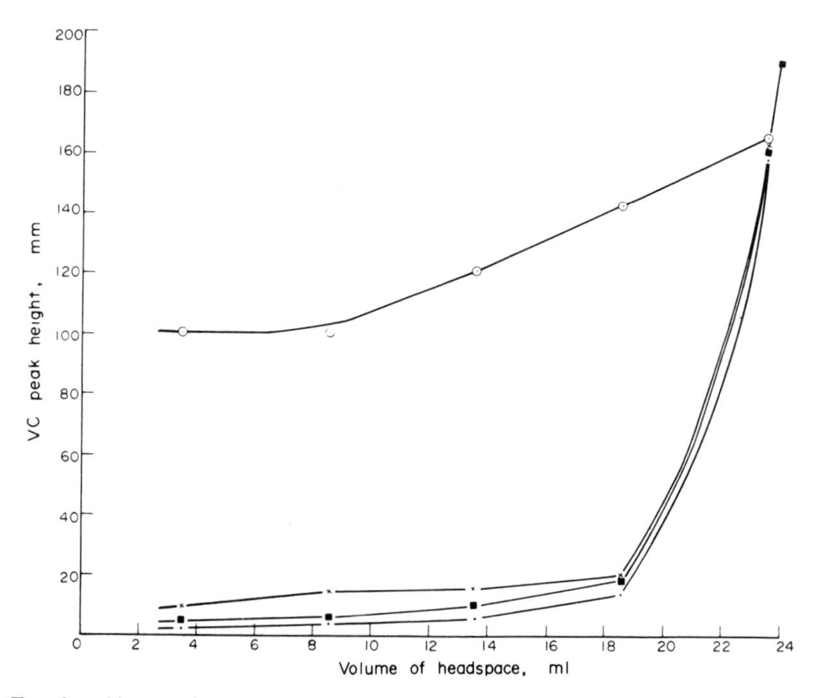

FIG. 3. Changes in the VCM content of headspace gases at different ratios of headspace volume to VCM solution volume using water (—⊙—), MEK (—×—), PBS (—■—) and DMA (—·—).

including organic solvents, enabling greater sensitivity and lower limits of detection to be attained.

One other advantage is that aqueous standards of VCM can be readily checked by an independent determination, e.g. titration. VCM in aqueous solution is estimated by reaction with potassium bromide/bromate solution. After bromination for several hours (preferably overnight in the dark) the excess bromine is determined by using potassium iodide and a titration of the liberated iodine with sodium thiosulphate using starch as the indicator.[2] The reaction is shown below:

$$CH_2{=}CHCl + Br_2 \rightarrow CH_2BrCHClBr$$

and

$$1\ \text{ml}\ 0{\cdot}1\text{N}\ Na_2S_2O_3\ \text{solution} \equiv 0{\cdot}003\,125\ \text{g VCM}$$

Organic solvents have also been widely used in recent years for VCM determinations and such standards are more stable and easier to handle, since VCM is more soluble in most organic solvents than in water. The choice of solvent is dependent mainly on the availability of a grade of sufficient purity, so that no interfering peaks are observed on the chromatogram in the areas of interest, viz. the VCM peak or the internal standard peak. The boiling point of the solvent should be as high as possible and the solvent should dissolve the plastic to form a mobile solution at about the 5 per cent level. Tetrahydrofuran can be used for PVC up to the 10 per cent level, but purification may be necessary, and the temperature of the chromatograph must be kept below 120 °C to prevent thermal decomposition. Ethanol, methylethylketone and dimethylacetamide are acceptable alternatives. The latter solvent is preferred by many workers in Europe because it elutes very slowly from the chromatographic column, and several analyses can be performed without interference. When peaks do emerge, the column has to be baked until clear, or alternatively, back flushing facilities can be incorporated.

The technique employed to prepare a standard solution is to add a known weight of solvent to the container, then bubble VCM through the solution and reweigh. Alternatively, VCM can be frozen and added in the liquid state. It is assumed that during the bubbling no solvent is lost by evaporation. This is certainly true for dimethylacetamide (b.pt. 165 °C) where the VCM is bubbled through for less than 1 min. Unfortunately it is not possible to check the strength of such standard solutions independently. However, with care it is possible to reproduce the concentration levels obtained to around 1 per cent. Differences should certainly be no greater than 5 per cent. Working solutions can then be prepared by simple dilution of the stock standards and, by comparison with aqueous standards, are less liable to losses of VCM during normal analytical operations. However, since the partition of VCM into the headspace above an organic solvent is much lower, the sensitivity achieved will be less than when working with aqueous systems.

Sampling

The difficulty of sampling effluents has already been referred to. Equally, the taking of a representative portion from a large PVC bottle is not as simple as it may appear. Whilst monomer levels in finished PVC plastics are now so low, little variation will be detected. However, some indication of the problem can be found by examination of Table 2, which shows the results obtained on analysis of samples taken from various parts of a one litre bottle (Fig. 4), of the type used for soft drinks. Some correlation between polymer thickness and residual monomer content is observed.

Whilst solid samples can be examined directly by headspace sampling techniques, they are more commonly analysed following solution or dispersion in a suitable solvent. This results from the difficulty in removing

TABLE 2
VARIATION IN VCM CONCENTRATION AT DIFFERENT SITES OF A PVC BOTTLE

Bottle	Site[a]	Polymer thickness (mm)	VCM concentration (μg/g)
1.	A	18	1·3
	B	16	1·1
	C	11	0·5
	D	18	1·3
	E	19	3·8
2.	A	99	1·1
	B	88	0·5
	C	88	0·4
	D	106	1·3
	E	107	4·0

[a] See Fig. 4 for an explanation of the sites.

the last traces of VCM from the middle of particulate matter without excessively long equilibration times. Berens *et al.*[38] have examined this problem and found that at a temperature above the glass transition point of PVC, the equilibration time falls to less than 60 min for particles of 25 μm or less in diameter, and Henry's law is obeyed up to levels of VCM as high as 4000 ppm. Moulded, extruded or fused samples of PVC were less satisfactory and much longer equilibration times were needed. However, the technique does have the advantage that as no solvent is required there is no dilution of the VCM in the sample and no time is lost in elution of solvent peaks from the chromatographic column.

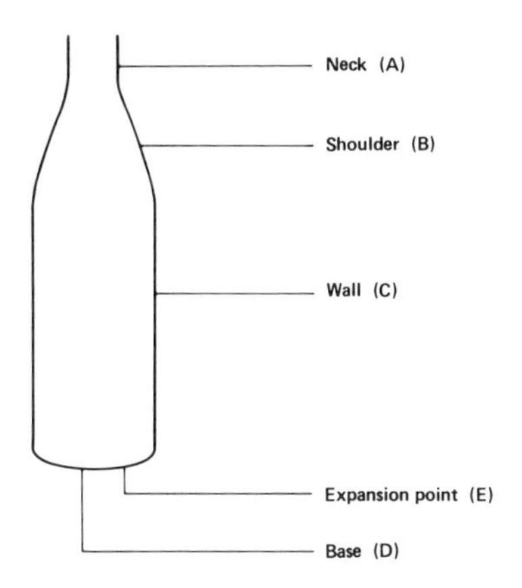

FIG. 4. Sampling positions on a PVC bottle.

Whilst with polymer samples solution techniques are readily employed, with foods it is not possible to dissolve the food in this way and to ensure that all VCM is released from particulate matter in the sample matrix. However, analysis of a number of different foods suggests that the problem is unlikely to lead to significant errors at the levels of VCM encountered. For this reason many workers prefer to use no solvent at all since the addition of an organic solvent as a dispersing agent will lower the concentration of VCM in the headspace and, hence, the detection limit attainable.

The above discussion is drawn largely from our experience in the determination of VCM in polymers and in foods over recent years. However, a number of other workers have published analytical methods and their findings will now be discussed. Williams and Miles[39] used a direct injection method for aqueous samples and a headspace method for samples of peanut oil. Three columns of different polarities were used for confirmation of identity, although mass spectrometry was available for levels greater than 50 ppb. A detection limit of 10 to 15 ppb was attained by direct injection and 5–10 ppb by headspace sampling. Small amounts of VCM were detected in alcoholic beverages, vinegars and peanut oil. In a later publication,[40] the headspace technique was also applied to samples of vinegar and beverages. The mass spectrometer confirmation by single ion monitoring at m/e 62 had been extended down to 10 ppb. The headspace of 'spiked' samples was found to contain 65 per cent of added VCM in vinegar, 60 per cent in sherry and 50 per cent for gin and martini, thus confirming the effects of different sample matrices. Direct injection and headspace methods were shown to give equivalent results. Similar studies on plastics, foods and simulating solvents have been reported.[41-45] In addition to practical descriptions of GLC procedures, these papers contain hints on the preparation and handling of standard solutions, results obtained following analysis of samples, as well as migration data. Generally, most methods based on headspace analysis can be used to detect levels of VCM down to 100 ppb in polymers and down to the range 1 to 5 ppb in foodstuffs. These levels, are adequate for current and future needs.

Dennison *et al.*[45] have designed a sparging apparatus in which VCM is purged from a large volume of the sample in solution, using helium for about 2 h. The VCM is trapped in ethanol cooled by a dry ice/solvent mixture. This pre-concentration step enables even lower levels of VCM to be detected and some interferences to be removed. A similar system was developed in these laboratories some years ago but the increase in sensitivity obtained was only marginal by comparison with the additional analysis time required.

RESULTS

The Steering Group on Food Surveillance sponsored by the Ministry of Agriculture, Fisheries and Food established a Working Party on Vinyl Chloride in 1973. This working party made up almost entirely of representatives of UK industry has reported[46] the levels of VCM found in bottles, rigid film and some foods in the period 1974–77. A summary of some of these results is presented in Table 3. It is clear that there has been a marked reduction in the VCM content of PVC manufactured articles during the period 1974–76, and this change is also reflected in the levels found in foods such as fruit drink, cooking oil, butter and margarine. Similar trends in the level of VCM in foods have been reported in other

TABLE 3

RESIDUAL LEVELS OF VINYL CHLORIDE MONOMER IN PVC PRODUCTS[46]

Stage of production	VCM content (ppm) 1974	1975	1976
Bottles: Level in polymer at time of use	500–1 000	100–250	15–50
Level in powder blend (maximum)	100	5	1
Level in bottle wall	100	5	1
Foil: Maximum level	150	15	5
Flexible film (extrusion blown) maximum	1	1	1

countries too as a result of modifications to the method of manufacture and improved blending and fabrication techniques. Levels in food are determined primarily by the initial VCM content in the PVC, together with storage time and temperature. The effect of these and other parameters on the migration of VCM from PVC articles into food is discussed more fully in Chapter 7.

OTHER MONOMERS

Problems encountered in the application of the headspace technique to the determination of other monomers are very similar to those already described for VCM. The main differences arise from the higher boiling points of the other monomers which have to be equilibrated at higher temperatures. Some change in the chromatographic separation conditions have also to be employed.

Vinylidene Chloride (VDC)

The principal use of VDC is as a copolymer with VCM in films. The presence of an additional halogen atom in the molecule increases the sensitivity of the electron-capture detector. Birkel *et al.*[47] described a method for the determination of VDC based on gas–solid chromatography using an ECD. Confirmation by mass spectrometry was also possible and levels of VDC down to 5 ppm could be detected. Recoveries of VDC added to solvent were almost quantitative and production runs of Saran film (Dow Chemical Co.) were found to contain from 6 to 30 ppm of VDC.

A similar technique was used by Hollifield and McNeal[48] for measurements of VDC in both food packaging films and in food simulating solvents using headspace sampling. Detection limits as low as 1 ppm in the polymer film and 5 to 10 ppb in food simulants were achieved. Lower levels of residual VDC were found in Saran film. In both methods, heptane and tetrahydrofuran were used as solvents in the preparation of standard solutions of VDC. Dilute working solutions need to be freshly prepared each day. THF gave rise to more troublesome interferences when using an ECD. Van Lierop *et al.*[32] used *m*-xylene as the solvent for VDC and with the Hall detector they were able to detect 50 ppb in aqueous foods and beverages. A later method[49] by Van Lierop used headspace sampling techniques and was applicable to all types of sample. Mass spectrometry in which the peak *m*/*e* 61 was selected, extended the method down to 5 ppb. Where interferences were suspected, other fragments such as *m*/*e* 63, 96 or 98 were examined. Methods for the detection of VDC in air have been described by Severs and Skory[50] and also by Russell.[51]

Acrylonitrile (AN)

AN is a component of several polymers used as food packaging materials. The basic material may contain as much as 70 per cent AN in conjunction with styrene and/or butadiene. Originally, methods for AN were based on polarography, following an azeotropic distillation with methanol.[52] Quantities of AN down to 0·1 ppm could be detected in industrial effluents. This method was extended by Crompton[53] to liquids obtained from extractability tests on various styrene–acrylonitrile copolymers. The presence of styrene monomer in the test solutions did not cause any interference up to levels of 500 ppm and AN could be measured down to 1 ppm with recoveries around 90 per cent.

Headspace sampling and analysis for AN has been described by Pasquale *et al.*[54] Dimethylformamide (DMF) was used as solvent for the resin when a FID was used. With the nitrogen-selective detector, dimethyl sulphoxide was the preferred solvent. Propionitrile was recommended as the internal standard and dilution of the resin dispersion with water as recommended by Steichen[36] was found to increase the sensitivity so that levels of AN down to 30 ppb could be detected in olive oil. Brown *et al.*[55] also used the nitrogen-selective detector along with gas–solid chromatography and were able to detect residual levels of 0·5 ppm of AN in the polymer. They point out the necessity of restricting the headspace volume above standard and test solutions to a minimum because of the volatility of AN. Furthermore, acetic acid was added to all solutions to condition the gas chromatographic column and so prevent excessive tailing of the AN peak. In a more recent paper[56] by the same group of workers, confirmation by mass spectrometry was described using peaks with m/e values of 51, 52, 53 and 26, normalised to m/e 53.

Peak tailing and the use of acetic acid for conditioning the column can be avoided by gas–liquid chromatography. Gawell[57] has described such a method, which is applicable to both packaging materials and to beverages. Using headspace sampling and the nitrogen-selective detector, AN down to 0·1 ppm in plastics and 5 ppb in beverages could be measured. Two chromatographic columns were used to confirm the identification of the AN peak and propionitrile was again used as the internal standard. Residual AN in Barex plastic was detected in the range 2 to 5 ppm and trace quantities were also found in some samples of beer and soft drinks packed in this product.

Some problems in the determination of AN have been reported by Chudy and Crosby.[37] They found that during the establishment of equilibrium in the headspace, differences were observed between water or DMF on the one

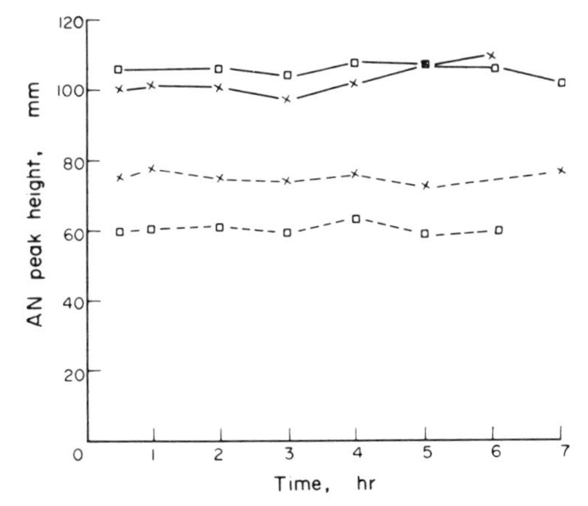

FIG. 5. The effect of temperature on changes in the AN concentration of headspace gases with storage time, using water at 60 °C (···□···) and 80 °C (—□—), and DMF at 60 °C (··· × ···) and 80 °C (— × —).

hand and DMA on the other. At 60 or 80 °C, equilibrium in water or DMF was established in less than 30 min and no further change was observed over 7 h (Fig. 5). However, when DMA was used as the solvent at 80 or 90 °C, the concentration of AN in the headspace fell rapidly after about 1 to 2 h as shown in Fig. 6. Furthermore, the disappearance of the AN peak on the chromatogram was matched by the simultaneous appearance of a new peak (NP) on the same chromatogram as shown in Fig. 7. The full explanation of

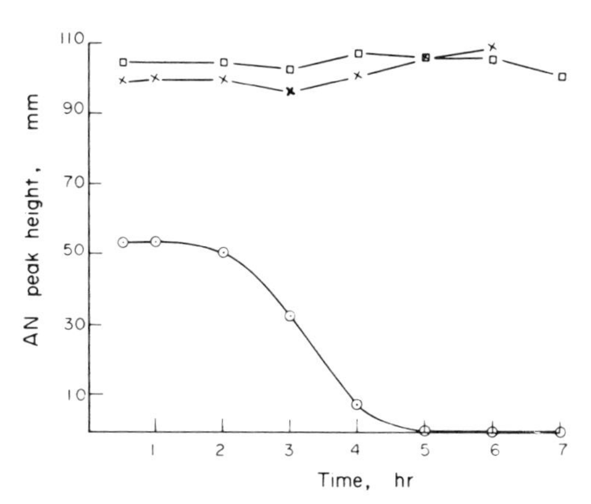

FIG. 6. The effect of solvent on changes in the AN concentration of headspace gases with storage time at 80 °C using water (—□—), DMF (— × —) or DMA (—⊙—).

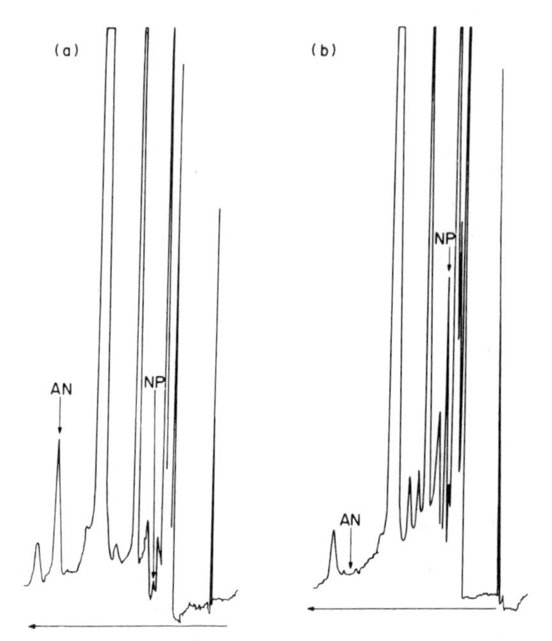

FIG. 7. Headspace gas chromatogram showing AN peaks using DMA as solvent at 80 °C (a) after 30 min equilibration and (b) after 7 h.

this phenomenon is not known, since it has only been observed in the one solvent and may be caused by impurities in that solvent. One possibility is that at high temperatures, above the boiling point, polymerisation reactions occur. However, since the new peak has a shorter retention time than AN, this seems to be an unlikely explanation.

Various gas chromatographic packings have been evaluated[51] for the adsorption of air pollutants, including acrylonitrile. Concentrations down to 1 ppm can be detected. A colorimetric method sensitive as low as 20 ppm has been adopted by IUPAC.[58] Alternatively, Roy[59] has described a titration method for the determination of residual monomers in styrene–acrylonitrile copolymers at the ppm level. Acrylonitrile is reacted with dodecanethiol and the excess of this reagent is estimated titrimetrically using silver nitrate solution.

Styrene

Styrene in the atmosphere may be determined by a colorimetric method[60] down to concentrations of around 200 ppm. Alternatively as for VCM, the monomer can be adsorbed onto charcoal, extracted with carbon disulphide and determined by gas–liquid chromatography.[61] An 85 to 90 per cent efficiency has been obtained in the range 50–200 ppm using this procedure. Dimethylformamide can also be used for extraction. Much lower concentrations (0·4 ppb to 0·4 ppm) have been detected[62] by adsorption onto charcoal, heat desorption and GLC. Conversion to the dibromide derivative and determination using an ECD is claimed[63] to be much more sensitive than using an FID. GC–MS has been used[64] to detect styrene in volatiles from a rubber vulcanisation plant. GLC has also been used[65] to detect low levels of styrene in effluents.

Residual levels of styrene in polymers can be determined titrimetrically after bromination as described by Roy,[59] or by gas chromatography as in British Standard BS: 2782: Part 4: Method 453A: 1978 (ISO: 2561-1974). In the latter procedure, the polymer is dissolved in chloroform, or dichloromethane, and then precipitated by the addition of methanol. A small volume of the supernatant is injected onto the GLC column with *n*-butylbenzene as the internal standard. The limit of detection is 10 ppm. Steichen[36] developed a headspace method in which water is used as the displacing solvent, giving a lower limit of detection of 1 ppm.

Methods used to detect styrene in foods have again been based on GLC. Boidron and Pons[66] developed a method for wine and claimed to be able to detect 0·005 ppm of styrene. Concentrations of styrene above 0·12 ppm were found to impart a characteristic taint to the wine. Withey and Collins[67,68] used headspace GLC to examine a wide range of Canadian foods including yoghurt, milk, butter, cream, cottage cheese and honey. They were able to detect 1 ppm in the polymer and down to 1 ppb or below in some samples of food. The detection limit was much higher for margarine owing to the unfavourable partition of styrene between the fat and the headspace. Davies[69] has developed methods based on GC–MS for the determination of styrene in foods such as orange squash, margarine and cream.

REFERENCES

1. EGAN, H., SQUIRRELL, D. C. M. and THAIN, W. (Eds.), *Environmental carcinogens—selected methods of analysis*, Vol. 2, *vinyl chloride*, International Agency for Research on Cancer, Lyon, 1979.
2. THAIN, W. (Ed.), *The determination of vinyl chloride; a plant manual*, Third edn, Chemical Industries Association Limited, London, 1978.
3. LAO, R. C., THOMAS, R. S. and MONKMAN, J. L., Improved methods for sampling and analysis of vinyl chloride, *Amer. Ind. Hyg. Assoc. J.*, 1976, **37,** 1.

4. McCLENNY, W. A., MARTIN, B. E., BAUMGARDNER, R. E., STEVENS, R. K. and O'KEEFE, A. E., Detection of vinyl chloride and related compounds by a gas chromatographic chemiluminescence technique, *Environ. Sci. & Technol.*, 1976, **10,** 810.
5. FREUND, S. M. and SWEGER, D. M., Vinyl chloride detection using CO and CO_2 infrared lasers, *Analyt. Chem.*, 1975, **47,** 930.
6. DENENBERG, B. A., MILLER, R. W. and KRIESEL, R. S., Instrumentation for continuous VCM monitoring, *Amer. Lab.*, 1975, **7,** 49.
7. HILL, R. H., McCAMMON, C. S., SAALWAECHTER, A. T., TEASS, A. W. and WOODFIN, W. J., Gas chromatographic determination of vinyl chloride in air samples collected on charcoal, *Analyt. Chem.*, 1976, **48,** 1395.
8. CUDDEBACK, J. E., BURG, W. R. and BIRCH, S. R., Performance of charcoal tubes in the determination of vinyl chloride, *Environ. Sci. & Technol.*, 1975, **91,** 1168.
9. PURCELL, J. E., Gas chromatographic analysis of vinyl chloride, *Amer. Lab.*, 1975, **7,** 99.
10. MILLER, B., KANE, P. O., ROBINSON, D. B. and WHITTINGHAM, P. J., Determination of a 24-hour time-weighted average value of environmental VCM concentration by charcoal adsorption followed by gas chromatography, *Analyst*, 1978, **103,** 1165.
11. AHLSTROM, D. H., KILGOUR, R. J. and LIEBMAN, S. A., Trace determination of vinyl chloride by a concentration/gas chromatography system, *Analyt. Chem.*, 1975, **47,** 1411.
12. BERTSCH, W., CHANG, R. C. and ZLATKIS, A., The determination of organic volatiles in air pollution studies: characterization of profiles, *J. Chromat. Sci.*, 1974, **12,** 175.
13. IVES, N. F., Sensitive trapping and gas chromatography method for vinyl chloride in air samples, *J. Assoc. Offic. Analyt. Chem.*, 1975, **58,** 457.
14. ASH, R. M. and LYNCH, J. R., Evaluation of gas detector tube systems: sulphur dioxide, *Amer. Ind. Hyg. Assoc. J.*, 1971, **32,** 490.
15. RAVEY, M. and KLOPSTOCK, J., Trace analysis of vinyl chloride in PVC and in the atmosphere, *J. Chromat. Sci.*, 1975, **13,** 552.
16. SHOU-YIEN HO, J., Collaborative study of reference vinyl chloride charcoal tubes, *Amer. Ind. Hyg. Assoc. J.*, 1979, **40,** 200.
17. LANDE, S. S., Measurement of atmospheric vinyl chloride, *Amer. Ind. Hyg. Assoc. J.*, 1979, **40,** 96.
18. WEST, P. W. and REISZNER, K. D., Field tests of a permeation-type personal monitor for vinyl chloride, *Amer. Ind. Hyg. Assoc. J.*, 1978, **39,** 645.
19. BELLAR, T. A., LICHTENBERG, J. J. and EICHELBERGER, J. W., Determination of vinyl chloride at μg/l level in water by gas chromatography, *Environ. Sci. & Technol.*, 1976, **10,** 926.
20. DRESSMAN, R. S. and McFARREN, E. F., A sample-bottle purging method for the determination of VC in water at submicrogram per liter levels, *J. Chromat. Sci.*, 1977, **15,** 69.
21. KOLB, B., Application of gas chromatographic head-space analysis for the characterisation of non-ideal solutions by scanning the total concentration range, *J. Chromatog.*, 1975, **112,** 287.
22. DROZD, J. and NOVÁK, J., Headspace determination of benzene in gas–aqueous

liquid systems by the standard additions method, *J. Chromatog.*, 1978, **152,** 55.

23. DROZD, J. and NOVÁK, J., Quantitative head-space gas analysis by the standard additions method. Determination of hydrophilic solutes in equilibrated gas–aqueous liquid systems, *J. Chromatog.*, 1977, **136,** 37.
24. HACHENBERG, H. and SCHMIDT, A. P. (Eds.), *Gas chromatographic headspace analysis* (Trans. Verdin, D.), Heyden, London, 1977.
25. BLADES, A. T., The flame ionization detector, *J. Chromatog. Sci.*, 1973, **11,** 251.
26. GIUFFRIDA, L., A flame ionization detector highly selective and sensitive to phosphorus, *J. Assoc. Offic. Agric. Chem.*, 1964, **47,** 293.
27. AUE, W. A., GEHRKE, C. W., TINDLE, R. C., STALLING, D. L. and RUYLE, C. D., Application of the alkali-flame detector to nitrogen containing compounds, *J. Gas Chromatog.*, 1967, **5,** 381.
28. BRAZHNIKOV, V. V., GUREV, M. V. and SAKODYNSKII, K. I., Thermionic detectors in gas chromatography, *Chromatog. Rev.*, 1970, **12,** 1.
29. COULSON, D. M., Electrolytic conductivity detector for gas chromatography, *J. Gas Chromatog.*, 1965, **3,** 134.
30. HALL, R. C., A highly sensitive and selective microelectrolytic conductivity detector for gas chromatography, *J. Chromatog. Sci.*, 1974, **12,** 152.
31. ERNST, G. F. and VAN LIEROP, J. B. H., Simple, sensitive determination and identification of vinyl chloride by gas chromatography with a Hall detector, *J. Chromatog.*, 1975, **109,** 430.
32. VAN LIEROP, J. B. H., HOGENDIJK, C. J. and JONGERIUS, TH. J., *De Ware (N)—Chemicus*, 1975, **5,** 158.
33. LOVELOCK, J. E., Gas chromatography 1968, *Proceedings of the 8th Internat. Symp. on Gas Chromatography*, Copenhagen, 1968, Harbourn, S. L. A. (Ed.), Institute of Petroleum, London, 1968, p. 95.
34. WILLIAMS, D. T., Semiquantitative method for confirming vinyl chloride monomer in foods, *J. Assoc. Offic. Analyt. Chem.*, 1976, **59,** 32.
35. HOFFMANN, D., PATRIANAKOS, C., BRUNNEMANN, K. D. and GORI, G. B., Chromatographic determination of vinyl chloride in tobacco smoke, *Analyt. Chem.*, 1976, **48,** 47.
36. STEICHEN, R. J., Modified solution approach for the gas chromatographic determination of residual monomers by head-space analysis, *Analyt. Chem.*, 1976, **48,** 1398.
37. CHUDY, J. C. and CROSBY, N. T., Some observations on the determination of monomer residues in foods, *Food Cosmet. Toxicol.*, 1977, **15,** 547.
38. BERENS, A. R., CRIDER, L. B., TOMANEK, C. J. and WHITNEY, J. M., Analysis for vinyl chloride in PVC powders by head-space gas chromatography, *J. Appl. Polym. Sci.*, 1975, **19,** 3169.
39. WILLIAMS, D. T. and MILES, W. F., Gas–liquid chromatographic determination of vinyl chloride in alcoholic beverages, vegetable oils and vinegars, *J. Assoc. Offic. Analyt. Chem.*, 1975, **58,** 272.
40. WILLIAMS, D. T., Gas–liquid chromatographic head-space method for vinyl chloride in vinegars and alcoholic beverages, *J. Assoc. Offic. Analyt. Chem.*, 1976, **59,** 30.
41. PU, H. H. and HICKS, S. Z., Determination of vinyl chloride monomer by gas chromatography, *J. Chromatog.*, 1977, **132,** 495.

42. DIACHENKO, G. W., BREDER, C. V., BROWN, M. E. and DENNISON, J. L., Gas–liquid chromatographic headspace technique for determination of vinyl chloride in corn oil and three food simulating solvents, *J. Assoc. Offic. Analyt. Chem.*, 1977, **60,** 570.
43. PAGE, B. D. and O'GRADY, R., Gas–solid chromatographic confirmation of VC levels in oils and vinegars by using electrolytic conductivity detection, *J. Assoc. Offic. Analyt. Chem.*, 1977, **60,** 576.
44. VAN LIEROP, J. B. H. and STEK, W., Analysis of vinyl chloride in food simulants at the low parts per billion levels by mass fragmentography, *J. Chromatog.*, 1976, **128,** 183.
45. DENNISON, J. L., BREDER, C. V., MCNEAL, F., SNIJDER, R. C., ROACH, J. A. and SPHON, J. A., Headspace sampling and gas–solid chromatographic determination and confirmation of ≥ 1 ppb VC residues in PVC food packaging, *J. Assoc. Offic. Analyt. Chem.*, 1978, **61,** 813.
46. HMSO Survey of Vinyl Chloride Content of PVC for Food Contact and of Foods, *Food Surveillance Paper No. 2*, Her Majesty's Stationery Office, London, 1978.
47. BIRKEL, T. J., ROACH, J. A. G. and SPHON, J. A., Determination of VDC in Saran films by EC/gas–solid chromatography and confirmation by MS, *J. Assoc. Offic. Analyt. Chem.*, 1977, **60,** 1210.
48. HOLLIFIELD, H. C. and MCNEAL, T., *J. Assoc. Offic. Analyt. Chem.*, 1978, **61,** 537.
49. VAN LIEROP, J. B. H., Personal communication.
50. SEVERS, L. W. and SKORY, L. K., Monitoring personnel exposure to vinyl chloride, vinylidene chloride and methyl chloride in an industrial work environment, *Amer. Ind. Hyg. Assoc. J.*, 1975, **39,** 669.
51. RUSSELL, J. W., Analysis of air pollutants using sampling tubes and gas chromatography, *Environ. Sci. & Technol.*, 1975, **9,** 1175.
52. DAUES, G. W. and HAMNER, W. F., Determination of small amounts of acrylonitrile in aqueous industrial streams, *Analyt. Chem.*, 1957, **29,** 1035.
53. CROMPTON, T. R., Determination of traces of AN in liquid extractants used in assessing the suitability of SAN co-polymers as food-packaging materials, *Analyst*, 1965, **90,** 165.
54. PASQUALE, G. D., IORIO, G. D. and CAPACCIOLI, T., *J. Chromatog.*, 1978, **160,** 133.
55. BROWN, M. E., BREDER, C. V. and MCNEAL, T., Gas–solid chromatographic procedures for determining AN in polymers and food simulating solvents, *J. Assoc. Offic. Analyt. Chem.*, 1978, **61,** 1383.
56. MCNEAL, T., BRUMLEY, W. C., BREDER, C. and SPHON, J. A., Gas–solid chromatographic–MS confirmation of low levels of AN after distillation from food simulating solvents, *J. Assoc. Offic. Analyt. Chem.*, 1979, **62,** 41.
57. GAWELL, G. B.-M., Determination of AN in plastic packaging and beverages by HSGC, *Analyst*, 1979, **104,** 106.
58. GAGE, J. C., STRAFFORD, N. and TRUHAUT, R., *Methods for the determination of toxic substances in air-acrylonitrile, method 21*, International Union of Pure and Applied Chemistry, Butterworths, London, 1962.
59. ROY, S. S., Titrimetric determination of residual monomers in styrene–acrylonitrile copolymers, *Analyst*, 1977, **102,** 302.

60. GAGE, J. C., STRAFFORD, N. and TRUHAUT, R., *Methods for the determination of toxic substances in air, method II, styrene*, International Union of Pure and Applied Chemistry, Butterworths, London, 1962.
61. BURNETT, R. D., Evaluation of charcoal sampling tubes, *Amer. Ind. Hyg. Assoc. J.*, 1976, **37,** 37.
62. PARKES, D. G., GANZ, C. R., POLINSKY, A. and SCHULZE, J., A simple gas chromatographic method for the analysis of trace organics in ambient air, *Amer. Ind. Hyg. Assoc. J.*, 1976, **37,** 165.
63. HOSHIKA, Y., Gas chromatographic determination of styrene as its dibromide, *J. Chromatog.*, 1977, **136,** 95.
64. RAPPAPORT, S. M. and FRASER, D. A., Gas-chromatographic mass spectrophotometric identification of volatiles released from a rubber stock during simulated vulcanization, *Analyt. Chem.*, 1976, **48,** 476.
65. AUSTERN, B. M., DOBBS, R. A. and COHEN, J. M., Gas chromatographic determination of selected organic compounds added to waste water, *Environ. Sci. & Technol.*, 1975, **9,** 588.
66. BOIDRON, J-N. and PONS, M., Dosage du styrene dans les vins, *Ann. Fals. Exp. Chim.*, 1978, **71,** 369.
67. WITHEY, J. R. and COLLINS, P. G., Styrene monomer in foods. A limited Canadian survey, *Bull. Environ. Contam. Toxicol.*, 1978, **19,** 86.
68. WITHEY, J. R., Quantitative analysis of styrene monomer in polystyrene and foods including some preliminary studies of the uptake and pharmacodynamics of the monomer in rats, *Environ. Health Perspectives*, 1976, **17,** 125.
69. DAVIES, J. T., Migration of styrene monomer from packaging material into food. Experimental verification of a theoretical model, *J. Food Technol.*, 1974, **9,** 275.

Chapter 4

TOXICOLOGICAL ASPECTS

INTRODUCTION

All substances are poisons; there is none which is not a poison. The right dose differentiates a poison and a remedy.
PARACELSUS (1493–1541)

Toxicology is the study of the interaction between chemicals and biological (living) systems. Originally, such studies were mainly concerned with the short-term (or acute) effects of poisonous compounds. More recently, chronic (or long-term) effects resulting from the ingestion of low levels of such compounds over a long period of time, in relation to the life span of the animal under test, have been investigated. All compounds are toxic if taken in sufficient quantity. Even oxygen and water, whilst essential constituents for all living systems, can be dangerous in excess. Healthy adults can usually tolerate breathing pure oxygen for up to 3 h without exhibiting any uncomfortable symptoms. However, further inhalation at atmospheric pressure (or short-term inhalation of high concentrations of oxygen at 2–3 atm) results in damage to the eye, e.g. bilateral progressive constriction of the peripheral fields, impaired central vision, mydriasis and constriction of the retinal vasculature.[1,2]

Not all interactions resulting from the presence of chemicals are harmful, particularly where the change is reversible. However, it is the harmful or deleterious and irreversible changes that can occur with which the toxicologist is mainly concerned. Whilst all substances can be considered to be toxic (as indicated above) it is usual to discriminate between non-toxic and toxic compounds on the basis of the dose required to produce a noticeable and deleterious effect. Thus, everyone would regard potassium

TABLE 1
MEDIAN LETHAL DOSE OF SOME COMMON SUBSTANCES ADMINISTERED ORALLY TO RATS

Compound	LD_{50} (mg/kg rat body weight)
Potassium cyanide	10
Dimethylnitrosamine	30
Lead tetraethyl	35
Acrylonitrile	80–90
Lead	100
Arsenic trioxide	140
DDT	400
Phenobarbital	660
Aspirin	1 500
Vinylidene chloride	~1 500
Table salt	3 000
Styrene	~5 000

cyanide as a toxic (poisonous) chemical, whilst few would classify sodium chloride as other than non-toxic, although patients suffering from hypertension are usually advised to consume a diet low in salt content.[3] Examples of median lethal dose levels for some commonly encountered chemicals are given in Table 1. Table 2 shows the classification of dangerous substances on the basis of LD_{50} values approved by the European Economic Community.

The purpose of toxicological studies and evaluations is to assess the extent of any hazard to human health arising from the use of chemicals in

TABLE 2
GENERAL CLASSIFICATION OF DANGEROUS SUBSTANCES

	LD_{50} absorbed orally in rat (mg/kg)	LD_{50} percutaneous absorption in rat or rabbit (mg/kg)	LD_{50} absorbed by inhalation in rat (mg/litre/4 h)
Very toxic	<25	<50	<0·5
Toxic	25–200	50–400	0·5–2
Harmful	200–2 000	400–2 000	2–20

Taken from the *Official Journal of the European Economic Community*, L 259 Vol. 22, 15 October, 1979.

food processing, as drugs, or from their occurrence in the environment, either naturally or as a result of industrial pollution. Not all hazards are man-made. Since earliest times civilisations have had to contend with natural catastrophes such as earthquakes, lightning, floods, drought and fires, as well as infectious diseases that could decimate whole communities in a very short time. Technology, whilst not an unmixed blessing, has certainly enriched human life in many ways. However, many hazards are undoubtedly technological in origin and have changed over the years as society has developed. Thus, malaria, typhoid, smallpox, tuberculosis and the vitamin-deficiency diseases have been largely conquered; only to be replaced by diseases of the heart, the lung, various cancers and problems caused by obesity. Equally, hazardous working conditions in factories and coal mines at the time of the industrial revolution, and the pollution of rivers at the turn of the last century when rivers were referred to as 'open sewers', are now much less severe. Adulteration of food and drugs has again been much reduced over the last 100 years. However, new problems have arisen through the use of a rapidly expanding chemical industry over a similar time scale. Examples of such new products include pesticides, medicinal compounds in animal feedingstuffs and for human use, petroleum products and propellants. As yet, many of these problems are not fully understood or precisely defined. Resolution of these difficulties can only be achieved through the concerted and combined efforts of industrialists, analytical chemists and toxicologists in particular.

RISK/BENEFIT ANALYSIS

Given that nothing can ever be made absolutely safe, it follows that one must evaluate the risk of any particular activity and, by balancing such risks against possible benefits of indulging in the activity, make a judgement as to whether the activity is worthwhile on the basis that the benefits outweigh any known and acceptable risks. Some common risks to life in the United Kingdom have been quantified by the Royal Commission on Environmental Pollution (Table 3), and in the United States of America by the National Center for Health Statistics (Table 4). Whilst some risks are avoidable by personal choice and life style (e.g. smoking), others are not so. Food consumption is essential for life and, hence, all activities are directed towards making the nation's food supply as safe as possible, based on existing knowledge of the mechanisms of the risks involved.

TABLE 3

PROBABILITY OF DEATH FOR AN INDIVIDUAL PER YEAR OF EXPOSURE

Risk	Activity
1/400	Smoking, 10 cigarettes per day
1/2 000	All accidents
1/8 000	Traffic accidents
1/20 000	Leukaemia from natural causes
1/30 000	Working in industry
1/30 000	Drowning
1/100 000	Poisoning
1/500 000	Natural disasters
1/1 000 000	Rock climbing for 90 s
1/1 000 000	Driving, 50 miles
1/2 000 000	Struck by lightning

Royal Commission on Environmental Pollution, Sixth Report, HMSO, 1976.

TABLE 4

SOME COMMON RISKS TO LIFE IN THE UNITED STATES

Cause of death	Number deaths/year	Number deaths/year/100 000 people
Heart disease	716 200	336
Cancer	365 700	173
Stroke	194 000	92
Auto accidents	45 900	22
Suicide	27 100	13
Drowning	8 000	3·8
Poisoning	6 300	3·0
Fire	6 100	2·9
Air travel	1 600	0·76
Railway accidents	600	0·28
Lightning	120	0·06
Cataclysm (e.g., tornado, flood, earthquake)	100	0·05

Based on 1975 data from the National Center for Health Statistics, Health Services and Mental Health Administration, US Department of Health, Education and Welfare.

The Frawley Concept[4]

In the particular case of food packaging materials, an attempt to assess quantitatively the available scientific evidence and determine the level of potential hazard was made by Frawley.[4] He started by an examination of all the published two-year chronic toxicity studies from which he tabulated a 'no-effect' level for each chemical. The distribution of these 'no-effect' levels, for 220 chemicals included in the survey, is shown in Table 5. A further division of the chemicals into two classes, namely, (a) heavy metals and pesticides, and (b) other compounds, showed that all nineteen compounds found to be toxic below 10 ppm were included in group (a). Of

TABLE 5

DISTRIBUTION OF NO-EFFECT LEVELS IN TWO-YEAR CHRONIC STUDIES

No-effect level (ppm)	All compounds (220)	Heavy metals and pesticides (88)	Others (132)
<1	5	5	0
<10	19	19	0
<100	40	39	1
<1 000	101	72	29
<10 000	151	86	65

Note: From work by J. P. Frawley.[4]

the 40 compounds found to be toxic at levels of 100 ppm or below, 39 were in group (a); the single exception being acrylamide. Since pesticides (and also many heavy metals formerly used for their biocidal properties) are designed specifically to be toxic to one or more forms of life in order to be successful commercially, this clear distinction between the two groups is not surprising. Furthermore, this analysis suggests that, excluding pesticides and heavy metals, fewer than 1 in 100 chemicals will have a 'no-effect' level below 100 ppm, and fewer still would be expected to exhibit toxicity at 10 ppm or below. Applying the usual (*vide infra*) 100-fold safety factor, it would appear that all of the 132 compounds in group (b) would be safe in the diet at a level of 0·1 ppm, without making any allowance for the difference in food intake between animals and man.

The second fundamental proposition in the Frawley concept concerns migration from the packaging material into foods. Frawley attempted to determine the maximum level of addition of a compound to a packaging material that would not be leached out into the food, at a level greater than

0·1 ppm. This will obviously vary from one material to another and from one food to another (see Chapter 7). For purposes of illustration, he chose paper as a substrate which was both highly permeable and subject to extraction by a variety of foods. Radioactively labelled rosin size was incorporated at three different levels into several typical commercial grades of paper. A wide range of foodstuffs (24 different types) were then packaged for different periods at different temperatures before examination. The parameters were selected to simulate extreme conditions of migration in the worst possible case. The average migration determined for each commodity group is shown in Table 6. From a knowledge of the contribution of each

TABLE 6

CALCULATION OF MAXIMUM MIGRATION OF ROSIN SIZE FROM PAPER INTO TOTAL DIET[4]

Commodity group	Per cent of diet	Average migration (ppm)	Contribution to total diet (ppm)
Milk products	31	3·1	1·0
Vegetables	20	2·0	0·4
Meats	18	38·2	6·9
Fruits	13	0·5	0·1
Grain products	10	3·5	0·4
Sugar	5	0·2	0·0
Butter, oils	3	32·8	0·9
Total			9·7

Note: 4 per cent size impregnated into paper.

commodity group to the total diet of an individual, it is possible to calculate the migration from each group into the diet as shown (Table 6), assuming that all foods are packaged in rosin sized paper. A good correlation between the level of size incorporated into the paper initially (in the range 1 to 4 per cent) and the migration found experimentally was established. Approximately 2 ppm for each percentage of size in the paper migrated into the diet. Assuming that no more than 25 per cent of man's diet is in contact with any one given type of food packaging material or additive, extrapolation of the above data suggests that a concentration of 0·2 per cent of an additive in a packaging material would result in a maximum concentration of 0·1 ppm of the additive in the diet. However, it is possible that the above conclusion is biased by the major contribution to total migration in the diet of the meat and butter commodity groups, i.e. fat-rich foods. Nevertheless, the experiment was designed to simulate an extreme case. In many instances migration would be negligible even at relatively high additive concentrations. The conclusion of the Frawley concept is, therefore, that any chemical suitable for use in food packaging is safe for man at a level in the diet not exceeding 0·1 ppm (or at a level of 0·2 per cent in the packaging material).

METHODS OF TOXICOLOGICAL TESTING

Since absolute safety is unattainable, methods of testing are designed to identify the extent of any possible potential hazard, or the conditions of use within which a substance is, to all intents and purposes, safe. Unfortunately, these methods take a long time, are very costly and require considerable experience in their application and in the interpretation of the results. The object of such testing is to gather information on such aspects as:

(1) acute toxicity
(2) sub-acute toxicity
(3) chronic toxicity
(4) metabolism and pharmacokinetics
(5) genetic response

The whole subject is undoubtedly complex and a wide range of chemical and pathological examination protocols have been developed over the years. The reader is referred to the bibliography at the end of this chapter for a more detailed and specialist presentation of this topic. Some of the most important points for a general understanding of the subject, together with some problem areas are discussed below.

Testing starts with the determination of acute toxicity based on the effect of a single dose (or repeated doses within 24 h) administered in the animal's feed or by injection. The dose–response relationship is fundamental to toxicology. Response may be difficult to define and measure unless based on an unequivocal reaction such as death (or the fraction of a population of animals to die). The fundamental dose–response relationship is illustrated in Fig. 1. The sigmoidal curve has a roughly linear portion between 20 and 80 per cent response and a correspondingly less reproducible response outside these limits. For this reason, Trevan[5] introduced the concept of LD_{50}, i.e. the lethal dose by which 50 per cent of the population of animals, of a given species, are killed. LD_{50} values are usually expressed in units of milligrams of compound per kilogram body weight (of the animal under

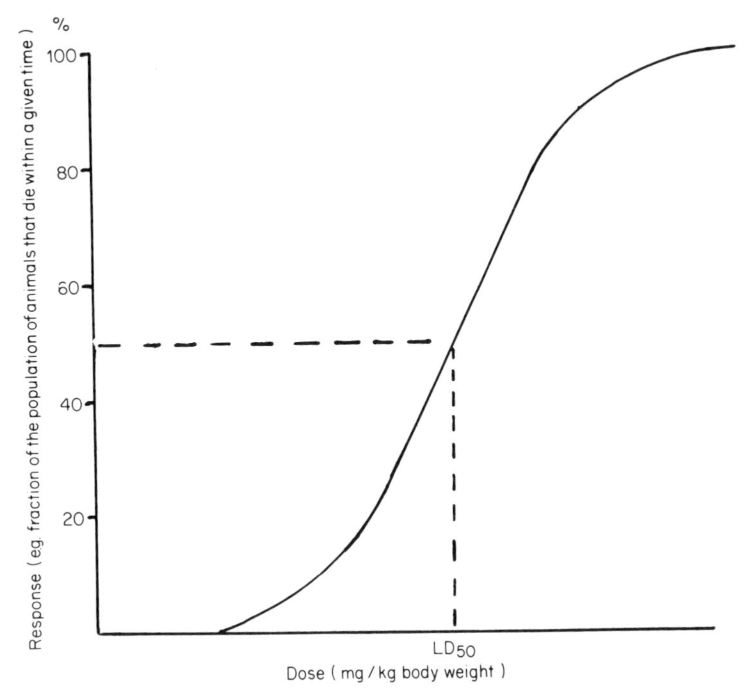

FIG. 1. Dose–response relationship in toxicity studies.

test), and are found to vary with such factors as species, route of administration, sex and nutritional status of the animals. Hence, experimental conditions must be fully defined. Nevertheless, the LD_{50} value is a quantitative, if somewhat crude, measure of toxicity. It permits comparisons to be made with other compounds previously tested and it gives some guidance as to suitable doses for use in later (chronic) testing, as well as an indication of the clinical effects to be observed. Highly toxic compounds (low LD_{50} values—Table 1) would be rejected at this stage but obviously no substance used as a direct or indirect additive to food falls into this category. Ideally, at least three species (including one non-rodent) should be tested as a check on species susceptibility. Both sexes should be used in at least one of the species tested.

Sub-acute experiments are designed to provide further knowledge of the biological effects and usually extend for a period of up to 10 per cent of the animal's lifetime. Information is obtained regarding the maximum tolerated dose, as well as the lowest dose required to produce an observable

TABLE 7
CHEMICAL CARCINOGENS

Compound	Source	Cancer site
Polynuclear aromatic hydrocarbons e.g. benzo[*a*]pyrene 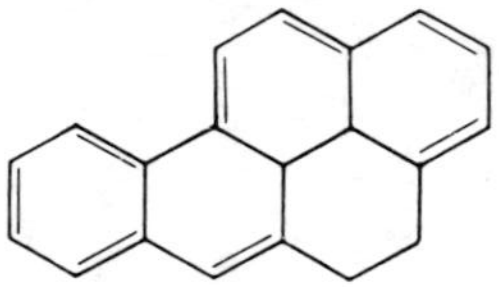	Mineral oils, soot, pitch, coal-tar, incomplete combustion of organic matter and smoked foods (?)	Skin
 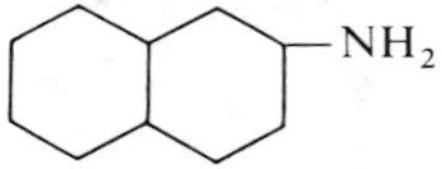 Aromatic amines e.g. benzidene (NH_2—, —NH_2)	Chemical and rubber industries, laboratory reagents	Bladder
2-naphthylamine (—NH_2)		

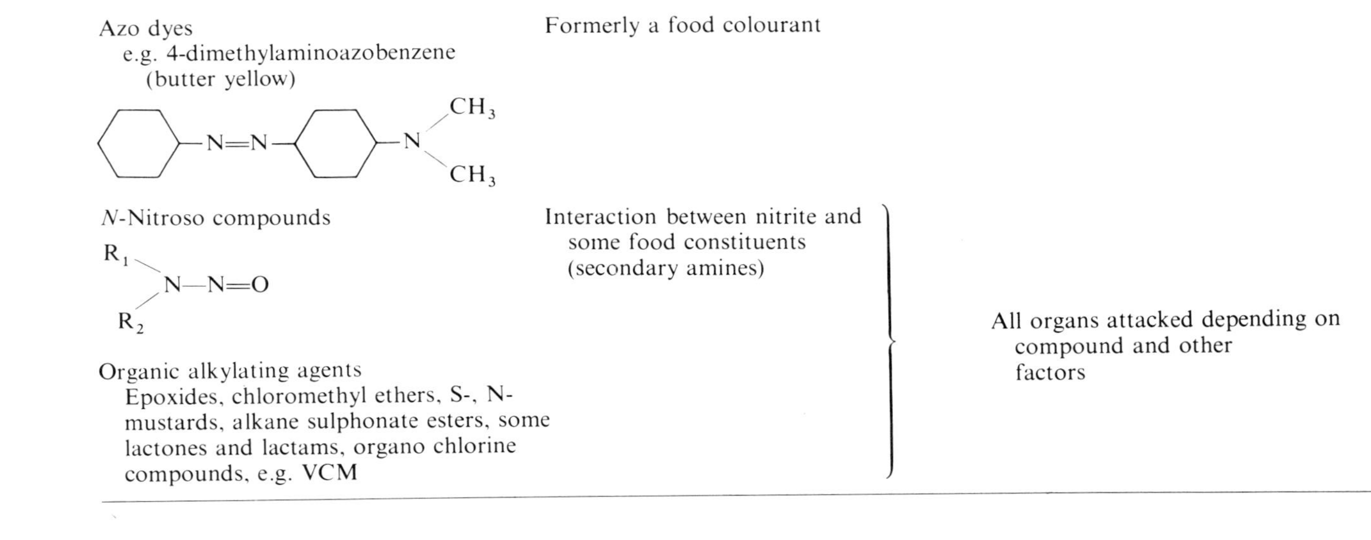

Azo dyes e.g. 4-dimethylaminoazobenzene (butter yellow)	Formerly a food colourant	
N-Nitroso compounds $R_1R_2N{-}N{=}O$	Interaction between nitrite and some food constituents (secondary amines)	All organs attacked depending on compound and other factors
Organic alkylating agents Epoxides, chloromethyl ethers, S-, N-mustards, alkane sulphonate esters, some lactones and lactams, organo chlorine compounds, e.g. VCM		

TABLE 8

SOME NATURALLY OCCURRING POTENTIAL CARCINOGENS IN FOOD

Compound and formula	Food	Effect
Mycotoxins, aflatoxins B_1, B_2, G_1, G_2 ochratoxin, patulin, penicillic acid sterigmatocystin, luteoskyrin 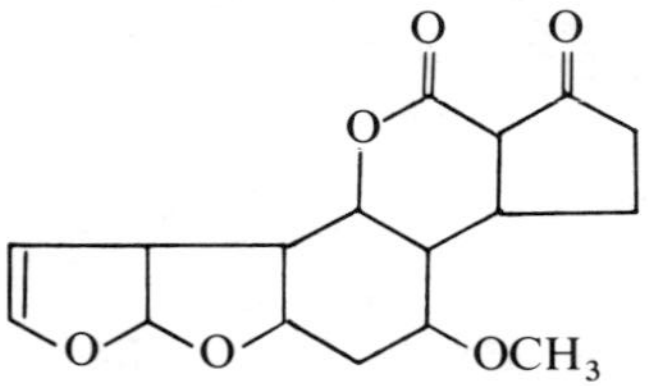 Aflatoxin B_1	Widely distributed amongst staple foods in hot climates, e.g. peanuts, beans, corn	Hepatocellular carcinogen, active in several species, some circumstantial evidence for humans (aflatoxins) Less complete data for other compounds listed
Coumarin 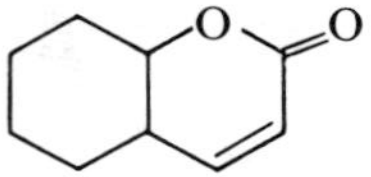	Flavourings, plants essential oils	Bile-duct carcinomas found in rats
Cycasins $CH_3—N(=O)=N—CH_2O\ C_6H_{11}O_5$	Seeds, roots and leaves of cycad plants found in the tropics	Carcinogenic in mice, rats, hamsters, guinea pigs, rabbits and fish—producing tumours in the liver, kidney and intestine

Safrole	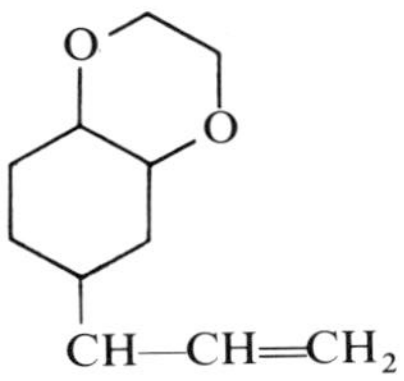 CH—CH=CH_2	Essential oils from nutmeg, mace, ginger, cinnamon and black pepper	Carcinogenic in mice and rats producing liver tumours
Tannins	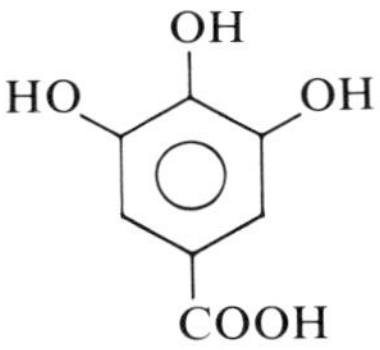 gallic acid; 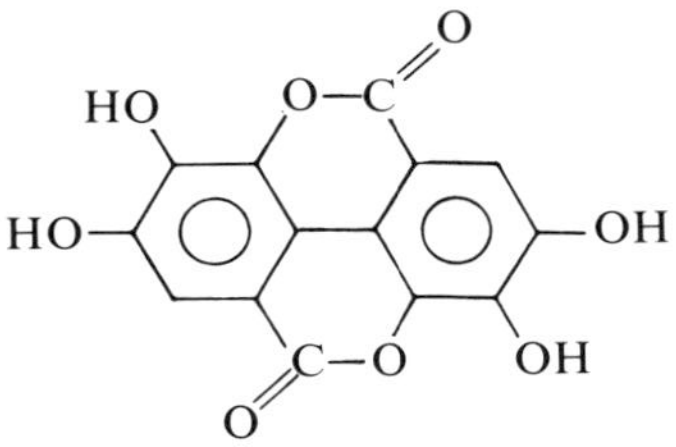 ellagic acid	Coffee, tea, used as flavouring and clarifying agents	Carcinogenic in rats following subcutaneous injection
Nitrosamines	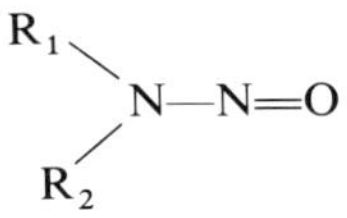 $R_1R_2N—N=O$	Trace quantities found in meat, fish, alcoholic beverages	Active in several species producing tumours of the liver and other organs

toxic effect. The mode of action must also be established through a study of absorption, excretion and metabolism of the compound within the animal's body, including any effects on enzyme systems in the blood and other tissues. For example, an increase in the level of certain enzymes such as the transaminases in the blood may be indicative of tissue damage. On the other hand, some substances produce harmful effects by enzyme inhibition. Ninety-day studies of compounds are frequently carried out by feeding low, medium and high doses to rats, during which time a number of clinical and biochemical measurements are made. Changes in body weight, food and water consumption and many other parameters are recorded during the lifetime of each animal. The lowest dose should produce no observable effects, whilst higher levels should be selected so that definite toxic effects can be seen. At the post-mortem examination, the tissues and organs are subjected to further histological and pathological tests. On occasions the classification of tumour-type, or morphological changes observed, can present problems and subsequently influence the result of statistical procedures. The observation of rare types of tumours, therefore, is usually less subject to challenge and of greater significance toxicologically. At least two species should be used and the number of animals in test and control groups should be large enough for the results to be statistically significant.

Carcinogenesis

Some 20 per cent of all deaths recorded annually in Great Britain,[6] and other countries too, are attributed to cancerous diseases. Environmental factors are thought to be a major cause of human cancer, and it is natural to suspect chemicals as being among the prime agents responsible, bearing in mind the changes observed over a period of time and from place to place, amongst different populations. Studies over a number of years have established the carcinogenicity of a number of chemical compounds. Benzo[*a*]pyrene was one of the first chemicals so identified by Cook *et al.*[7] in 1932 and since that time a number of other major groups of chemical substances have also been identified as carcinogens (Table 7). Well over 3000 chemical carcinogens have now been identified containing a variety of structural groupings. Carcinogenicity is not confined to man-made materials; indeed many such substances occur naturally (Table 8).

Chronic Toxicity

For food additives, the major interest centres round the possible effects of feeding very small quantities of the substance under test throughout the whole life-span of an animal—and the relevance of such effects to the human situation. Naturally, any such effects are less readily observable than in acute toxicity testing described above because of the lower dosing levels used. The relationship between short- and long-term feeding studies has been discussed by Weil and McCollister.[8] They attempted to predict the outcome of two-year chronic feeding trials on the basis of results obtained from 90-day tests. A quantitative comparison of the results for 33 materials showed that extrapolation from 90 days to two years could be made with confidence except for cholinesterase-inhibiting chemicals. Two-year feeding trials can also be conducted in which limited criteria only are evaluated. A minimum basic design was suggested which would enable a much larger number of compounds to be screened by elimination of the extensive microscopical examination of unaffected tissues. However, the supervision by an experienced toxicologist is essential for the final judgement and safety evaluation. Furthermore, even long-term studies are difficult to interpret since many of the animals in both test and control groups may die prematurely from natural causes despite recent improvements in animal husbandry. Prolongation of the life of animals also increases the natural incidence of tumours. Figure 2 shows the percentage tumour incidence for untreated rats that were allowed to live to the end of their natural life. So many tumours were observed after a span of 18–24 months that any increased effect caused by feeding a known compound would be very difficult to detect, since the response to a given dose is determined by the statistical difference between the number of tumours observed in test and control groups. For the differences to be statistically significant, large numbers of animals are required in both groups. Naturally, this increases the cost and the practical difficulties of testing. However, it is most important to ensure that any observed effects are indeed caused by feeding the substance under test and not by some other trace constituent present in the animal's food. Commercial diets need to be rigidly controlled, batch by batch, and examined for the presence of a wide range of contaminants, e.g. metals, insecticides, nitrosamines, polynuclear aromatic hydrocarbons, aflatoxins, drugs and any other chemicals known to be carcinogenic that may be present. Equally, the chemical under test must be clearly characterised and of known purity, although it may also be necessary to test commercial mixtures, in some cases where these are more closely related to the actual material, to which humans are exposed. In some countries, for example, powdered plastic is tested rather than the individual chemical compounds used in the formulation of the final plastic product. However, there is no guarantee (in this case) that the extractive properties of

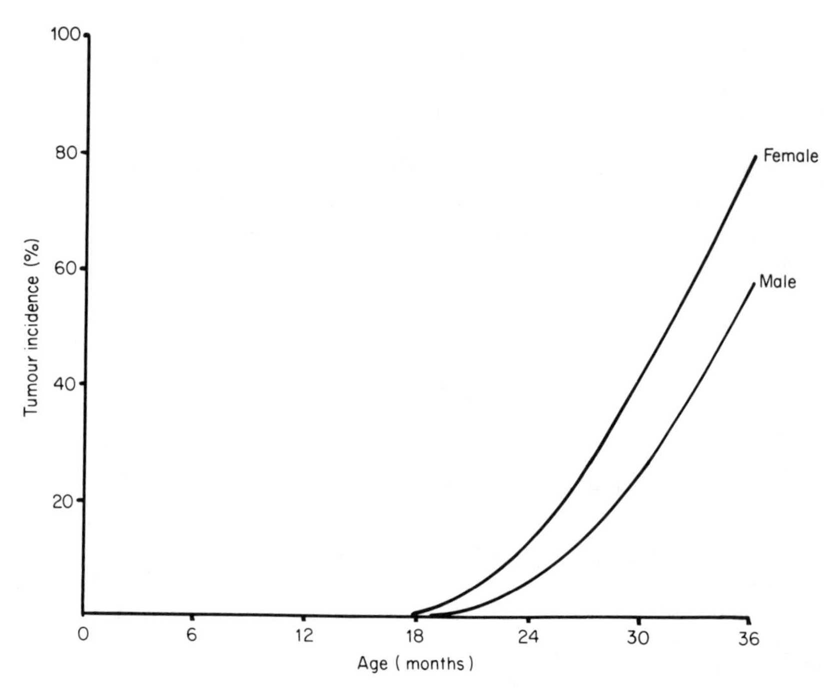

FIG. 2. Cumulative tumour incidence (per cent) with advancing age in untreated rats.[11]

the gut are in any way related to the migration that occurs during storage of food in such a plastic material.

Of fundamental importance in chronic toxicity and carcinogenicity studies is the establishment of a dose–response relationship for the compound. With increasing dose levels the percentage yield of tumours in a test group of animals will also increase. Equally, the induction period (time between the first dose and appearance of tumours) will decrease. The relationship has been described quantitatively by Druckrey and Schmähl[9] as follows:

$$d \times t^{2/3} = \text{constant}$$

where, d = the daily dose and t = the tumour induction period. Hence, at low dose levels the induction period will increase until, simultaneously, spontaneous tumours are observed in the control group of animals (Fig. 2). This makes it difficult, if not impossible, even for strong carcinogens, to extrapolate the experimental data to a theoretical no-effect level. A number of mathematical models have been described in which different assumptions are made relating dose to response. One of the best known is that proposed by Mantel and Bryan.[10] Statistical techniques were used to estimate a safe dose for carcinogens, given certain assumptions as to the risk levels considered to be acceptable. Mantel and Bryan[10] defined an arbitrary acceptable risk (1 in 100 million) and worked out a mathematical extrapolation from high (observed) dose levels to lower values with a statistical assurance level of 99 per cent. The slope of the extrapolation curve was very shallow giving a very conservative estimate of safe levels. This was an attempt to resolve the dilemma by which carcinogens are to be prohibited in foods at all levels above zero. Zero is defined here as the level that can be detected analytically. Obviously, for those compounds for which the analytical method is not particularly sensitive, the level permitted on such a basis could be potentially hazardous. Equally, where analytical methods can detect levels of 10^{-12} g, such legislative limits might be unduly restrictive. No doubt discussions as to the relative merits of these various definitions of 'safe' levels will continue for a long time. Concepts such as 'no-effect' level, zero tolerance, and no-threshold must be continuously re-examined in the light of the latest experimental data. It is instructive to calculate the number of molecules of a carcinogen (e.g. vinyl chloride) that are present in 1 kg of food at the 1 ppb level.

$$1 \text{ ppb} \equiv 1\ \mu\text{g/kg}$$

Hence, for vinyl chloride (molecular weight, 62·5)

$$\text{Number of molecules} = \frac{1 \times 10^{-6}}{62{\cdot}5} \times 6 \times 10^{23} \text{ (Avogadro's Number)}$$

$$\simeq 1 \times 10^{16}$$

With numbers as large as this, one wonders if the setting of legislative limits (or tolerances) for contaminants in food is as critical as is sometimes supposed, bearing in mind the toxicological significance at the molecular level of such a change in numbers as large as this (10^{16}).

'No-effect' levels are usually interpreted as, that level in the diet that gives rise to no *deleterious* effect in the animal. However, this is not always easy to determine. Grasso[11] has discussed the problem of non-specific injury caused by surface tension effects and hyperosmolarity, whereby even substances such as salt or glucose may give rise to tissue damage. Liver enlargement too, may be simply a result of the increased load on the animal's detoxification system or, alternatively, it may be a true manifestation of toxicity and pathological change.

Genetic Response

In addition to carcinogenicity testing, special studies to detect interference with the process of reproduction are often required. Such studies are carried out over two to three generations and are designed to produce information on fertility, mutagenicity and teratogenicity, as well as on the general progress of mothers and offspring. Mutations are generally undesirable and involve interference with DNA and subsequent replication. Tests have been designed to monitor both DNA damage and the repair of damaged DNA. Williams[12] has reviewed such test systems for the screening of chemical carcinogens. Since no single test is entirely satisfactory, a battery of bacterial mutagenic assays and DNA damage/repair tests are required for the screening of chemicals. Such tests are based on the hypothesis that carcinogenicity results from damage to DNA, the hereditary material of the cell. Mutations are also thought to be caused by DNA damage and, hence, chemicals that are found to be mutagenic are also likely to be carcinogenic. Organisms used in testing include fruit flies (*Drosophila*), yeasts, bacteria and mammalian cells. Obviously, there are difficulties inherent in the extrapolation of experimental data obtained using bacteria, yeasts or *Drosophila*, to possible hazards of using food additives in the human situation. Many studies have, therefore, been concerned with cell culture techniques from both rodent and human tissues and tumours *in vitro*. A summary of recent work has been published by Weinstein *et al.*[13] and by Loprieno.[14]

The best known screening test is undoubtedly that described by Ames,[15] which employs certain mutants of *Salmonella typhimurium* that have lost the facility to synthesise the amino acid histidine and cannot grow, therefore, in a culture medium which does not contain histidine. In the presence of chemicals possessing mutagenic activity, additional mutations can occur which result in repair of the DNA and the ability to grow in the absence of histidine. Mammalian liver tissue extracts are also incorporated into the test system to simulate metabolic changes to carcinogens that are activated *in vivo*. Using this test, some 300 chemicals have been tested and 90 per cent of all known carcinogens have been found to be mutagenic as well.[16] However, both false positives and false negatives have been reported. In contrast to animal feeding experiments, Ames tests (and other similar tests) are cheap and provide rapid results. They have much to offer as screening tests, but obviously cannot replace full biological testing and toxicological examination of prospective food additives found to be acceptable in the preliminary screen. In so far as they reduce the effort, time and skill that is largely wasted, particularly in the areas of clinical chemistry and histology during the 90-day test (Crampton[17]), mutagenic potential testing can make a significant contribution to safety evaluation procedures.

EXTRAPOLATION OF ANIMAL DATA TO MAN

Further uncertainties are introduced when animal data are extrapolated to the human situation in order to define an acceptable daily intake (ADI) for a given food additive in the national diet. This limit is based on the maximum tolerated dose obtained from animal studies, modified by a suitable safety factor. An arbitrary factor of 100 was recommended by the joint FAO/WHO Expert Committee on Food Additives[18] and this value has been widely accepted. In some instances there may be grounds for adopting a greater margin of safety. For example, when the additive is to be used in foods such as ice cream or soft drinks, which are consumed mainly by children, or when there is some doubt about the validity of the animal data. On occasions, lower safety margins can be defended. This may happen when extensive human data is available, where the compound is a natural body constituent, or when it is added to foods which make up a small part only of the national diet. In addition to the possible difference in sensitivity between the animal and man to a given compound, the safety factor also takes care of wide variations in sensitivity among the human population and between different age groups. It also provides a safety margin for those suffering from malnutrition or medical disorders who, therefore, might be more vulnerable than a normal healthy adult. Furthermore, test groups of animals are always very small in size compared to human populations so that even a 1 per cent increase in tumour incidence, even if statistically significant, could be catastrophic in human population terms. Animal data provide presumptive evidence only but, nevertheless, they do indicate the degree of concern necessary in the human situation. Chemicals known to be carcinogenic to man are also carcinogenic to animals, with the exception of benzene and arsenic.[19]

EPIDEMIOLOGICAL AND OTHER HUMAN STUDIES

Observations of the effects of a substance in man are of prime importance as a result of the known differences that occur between one species and another. Whilst direct experiments using possibly dangerous chemicals on humans are not normally possible, some information has been obtained through occupational studies and as a result of industrial accidents. The first occupational cancer to be recognised as such was 'chimney sweeps' disease, as described by Pott in 1775.[20] This followed from observations of a higher than average incidence of scrotal cancer compared to the population as a whole, helped by the appearance of the disease at a specific site in the body. In later studies, Cook *et al.* (1932)[7] established the carcinogenicity of certain polynuclear aromatic hydrocarbons, such as dibenz[*a*,*h*]anthracene and benzo[*a*]pyrene, in so far as they produced tumours after repeated painting of these materials onto the skin of experimental animals. Similar correlations have been postulated in the case of workers exposed to inadequately refined mineral oils, e.g. coal-tar workers, engineering operatives and mule spinners in the cotton industry. Polynuclear aromatic hydrocarbons have been identified in such products and in other situations where organic matter is incompletely burnt so that these chemicals are now widely dispersed throughout the environment. Many other chemical carcinogens have since been identified (Tables 7 and 8) and correlated with occupational diseases. Thus, cancer of the bladder among workers in the dyestuff and rubber industries has been attributed to contamination with aromatic amines.[21] Even inorganic compounds (especially asbestos) are known to present a potential industrial hazard. In contrast, the carcinogens so far identified in tobacco smoke cannot fully account for the incidence of lung cancer, without postulating a very high synergesis factor.

An alternative approach employs the study of cancer patterns in human populations throughout the world (epidemiology). Where an unusual or distinctive pattern is found, attempts are made to correlate the findings with an environmental factor, such as differences in the diet. Statistical methods are used to establish cause and effect relationships, by comparison with control groups. Thus, it is easy to correlate the increase in male and female lung cancer with the increase in the smoking of tobacco products following the first and second world wars respectively. However, epidemiological studies can only be used to make suggestions for further scientific studies. For example, the incidence of fluorosis led to the discovery of the beneficial effects of fluoridation, at controlled levels, on the reduction of dental caries.[22] Further difficulties in the epidemiological approach arise as a result of the lengthy latent period between initial exposure and tumour appearance. Furthermore, populations increasingly mix and interchange through increased travel and job mobility so that individuals are less likely to be exposed to a constant environmental factor for more than short periods of time. Nevertheless, evidence of human exposure is of enormous value when assessing safety levels. It may be possible to ignore apparent toxicological effects in animals when there is evidence of a long safety-in-use from human studies. Furthermore, if as a result of animal testing some food additive is to be banned, there may be other repercussions. For example, the current trend to reduce the number of permitted synthetic colours in foods has induced an increased body burden from the remaining permitted colours. Where there is doubt regarding the evidence of toxicity it may be better to encourage the use of a number of alternative compounds, so that the intake of any one given compound is reduced. In the case where the additive performs an essential safety function, e.g. the use of nitrates and nitrites in pork products as a check on botulism as well as for colour and flavouring, the prohibition of this additive raises the question as to how the public is to be protected from botulism in the future and whether the alternative means of control will prove more or less safe in years to come. Almost literally, in this case, 'jumping from the frying pan into the fire'.

The ever increasing use of chemicals in food has imposed upon public health authorities and governmental agencies the responsibility to decide: (a) on the technological need for a given chemical in foods; and (b) where the need has been established, to consider all aspects of the safety-in-use of such a chemical. This latter point can only be decided by specialists after the experimental observations described above. Consumer safety is then ensured through the enactment of legislation embodying positive lists of compounds that are permitted under prescribed conditions and negative lists of compounds that are to be prohibited on safety grounds. Further discussion of legislative aspects, relating to plastic materials can be found in Chapter 5. Some aspects of the toxicology of individual monomers will now be discussed.

EVALUATION OF THE CARCINOGENIC RISK OF SOME MONOMERS TO HUMANS

Vinyl Chloride

A detailed discussion of the relevant biological data and epidemiological studies on humans has been presented by an IARC Working Group.[23] Rats, mice and hamsters have all been exposed to varying levels of VCM, either by inhalation or by oral administration. Maltoni *et al.*[24] showed that inhalation of levels of VCM in the range 50 to 10 000 ppm in air for 4 h/day, 5 days/week for 12 months, resulted in the production of a number of tumours at different sites, including angiosarcomas of the liver. Oral administration of VCM in olive oil to rats was reported in a later study by the same group of workers.[25] Levels down to 3·33 mg/kg body weight were evaluated and some evidence was obtained of a dose–response relationship. Similar studies have been reported for mice and hamsters.[23] Vinyl chloride has also been shown to be mutagenic in the Ames test and metabolic studies provide some evidence of alkylation of nucleic acids. The principal products formed during metabolism are chloroethylene oxide and chloroacetaldehyde. Hopkins[26] has summarised the work on metabolism, macromolecular binding and carcinogenicity of vinyl chloride. Mutagenicity and metabolism of VCM have been reviewed by Bartsch *et al.*[27]

A wide range of toxic effects have been reported in human case studies.[23] The principal effects observed include, lesions of the bones in the terminal joints of the fingers and toes (acro-osteolysis) as well as changes in the liver and spleen. Long-term exposure gives rise to a rare form of liver cancer (angiosarcoma) and the association with exposure to VCM has been reported amongst plant operatives in several countries. In recent years, however, exposure to VCM at production and polymerisation plants has been markedly reduced[28] (Table 9) but the long latent period for tumour

TABLE 9
EXPOSURE LEVELS AT VINYL CHLORIDE PLANTS[28]

Year	Exposure level (ppm)
1945–55	1 000
1955–60	400–500
1960–70	300–400
Mid-1973	150
1975	5

development (20 years or more) may mean that further cases will come to light over the next decade. Histological evidence from 41 cases in the UK in the period 1963–73 was reviewed by a panel of pathologists, Baxter *et al.*[29] A diagnosis of death from angiosarcoma of the liver was confirmed in only 14 cases and in only one of these was there any history of significant exposure to VCM. However, the IARC Working Party did conclude that vinyl chloride causes angiosarcomas of the liver as well as tumours of the brain, lung and haematolymphopoietic systems in humans.[30]

Vinylidene Chloride

Less is known of the toxicology of VDC, both in animals and in humans. Whilst a number of studies are under way, few have yet been satisfactorily completed and the findings analysed. The LD_{50} value for rats is around 1500 mg/kg body weight, while in mice the value is 200 mg/kg body weight. The biological effects of VDC have been reviewed by Haley[31] and also by Cooper.[32] VDC affects the activity of several rat liver enzymes and decreases the store of glutathione. Some tumours have been observed after prolonged exposure but no teratogenic effects were seen in rats or rabbits. Different responses have been noticed between fed and fasted animals, and some species differences have been detected. The main pathway of excretion is via the lungs, with other metabolites being discharged by the kidneys.

Acrylonitrile

AN is considerably more toxic than the chlorinated monomers and has LD_{50} values of 80–90 mg/kg body weight in rats and 27 mg/kg body weight in mice. It has also been shown to be mutagenic after metabolic activation with liver enzymes. In animals AN is metabolised to cyanide, which is converted to thiocyanate and excreted in the urine.[33] There is also some evidence[23] of carcinogenicity in animals and possibly humans too, as a result of which the 8-h TWA is to be reduced to 2 ppm by 1981.

Styrene

The biological properties of styrene have been reviewed by Liebman.[34] The LD_{50} value for rats is 5 g/kg body weight. It is metabolised to styrene oxide which is a potent mutagen in a number of test systems. Both styrene and its oxide have been shown to produce chromosomal aberrations under certain conditions.[35] Further metabolism of styrene oxide produces hippuric acid. Toxic effects of styrene in humans have been reviewed by IARC.[23] The most frequently observed changes were of a neurological and psychological nature.

There is no doubt that our knowledge of the toxicity and carcinogenicity of monomers used in the production of plastics is far from complete, particularly as regards hazard to humans. Nevertheless, there is clear evidence of the need for concern in the case of vinyl chloride and possibly acrylonitrile too, whereas vinylidene chloride and styrene need only to be kept under review and the situation re-assessed as fresh evidence comes to light. However, such hazards relate solely to the presence of the monomer and not to the polymer itself where levels of residual monomer have been progressively reduced. Thus, any hazard which exists is directed chiefly towards workers in and around production plants, rather than the general population through the use of plastics as food packaging materials.

REFERENCES

1. Nichols, C. W. and Lambertsen, C. J., Effects of high oxygen pressures on the eye, *N. Engl. J. Med.*, 1969, **281,** 25.
2. Mailer, C. M., Paradoxical differences in retinal vessel diameters and the effect of inspired oxygen, *Can. J. Ophthalmol.*, 1970, **5,** 163.
3. Davidson, S., Passmore, R. and Brock, J. F., In: *Human nutrition and dietetics*, Churchill Livingstone, Edinburgh and London, 5th edn, 1972, p. 331.
4. Frawley, J. P., Scientific evidence and common sense as a basis for food-packaging regulations, *Food Cosmet. Toxicol.*, 1967, **5,** 293.
5. Trevan, J. W., The error of determination of toxicity, *Proc. Roy. Soc. B.*, 1927, **101,** 483.
6. Thomas, H. F., Cancer and the environment, *Environ. Health*, 1978, **86,** 232.
7. Cook, J. W., Hieger, I., Kennaway, E. L. and Mayneord, W. V., *Proc. Roy. Soc. B.*, 1932, **111,** 455.
8. Weil, C. S. and McCollister, D. D., Relationship between short- and long-term feeding studies in designing an effective toxicity test, *J. Agr. Food Chem.*, 1963, **11,** 486.
9. Druckrey, H. and Schmähl, D., *Naturwissenschaften*, 1962, **49,** 19.
10. Mantel, N. and Bryan, W. R., Safety testing of carcinogenic agents, *J. Natl. Cancer Inst.*, 1961, **27,** 455.
11. Grasso, P., Carcinogenic risks from food—real or imaginary? *Chemy. Ind.*, 1979, **3,** 73.
12. Williams, G. M., Review of *in vitro* test systems using DNA damage and repair for screening of chemical carcinogens, *J. Assoc. Off. Analyt. Chem.*, 1979, **62,** 857.
13. Weinstein, B. I., Grunberger, D., Fujimura, S. and Fink, L. M., Chemical carcinogens and RNA, *Cancer Res.*, 1971, **31,** 651.
14. Loprieno, N., Some problems associated with the testing for mutagenic potential, In: *Chemical toxicology of food*, Galli, C. L., Paoletti, R. and Vettorazzi, G. (Eds.), Elsevier, Oxford, 1978.
15. Ames, B. N., McCann, J. and Yamasaki, E., Methods for detecting carcinogens and mutagens with the *Salmonella*/mammalian-microsome mutagenicity test, *Mutat. Res.*, 1975, **31,** 347.
16. McCann, J., Choi, E., Yamasaki, E. and Ames, B. N., Detection of carcinogens as mutagens in the *Salmonella*/microsome test: Assay of 300 chemicals, *Proc. Nat. Acad. Sci. USA*, 1975, **72,** 5135.
17. Crampton, R. F., Current methodological approaches to the evaluation of chemical toxicity, In: *Chemical toxicology of food*, Galli, C. L., Paoletti, R. and Vettorazzi, G. (Eds.), Elsevier, Oxford, 1978.
18. Joint FAO/WHO Expert Committee on Food Additives, Second Report, 1958, *WHO Tech. Rep. Ser.*, No. 144.

19. Kraybill, H. F., Proper perspectives in extrapolation of experimental carcinogenesis data to humans, *Food Technol.*, 1978, **32,** 62.
20. Pott, P., *Chirurgical observations*, London, 1775.
21. Case, R. A. M., Hosker, M. E., McDonald, D. B. and Pearson, J. T., The role of aniline, benzidine, 1-naphthylamine and 2-naphthylamine, *Brit. J. Ind. Med.*, 1954, **11,** 75.
22. HMSO, The fluoridation studies in the UK and the results achieved after 11 years, *Reports in Public Health and Medical Subjects*, No. 122, HMSO, London, 1969.
23. IARC, *Monograph on the evaluation of the carcinogenic risk of chemicals to humans*, Vol. 19, International Agency for Research on Cancer, Lyon, 1979.
24. Maltoni, C., Lefemine, G., Chieco, P. and Carretti, D., Vinyl chloride carcinogenesis: current results and perspectives, *Med. Lav.*, 1974, **65,** 421.
25. Maltoni, C., Vinyl chloride carcinogenicity, In: *Origins of human cancer*, Hiatt, H. H., Watson, J. D. and Winsten, J. A. (Eds.), Cold Spring Harbor, New York, 1977, p. 119.
26. Hopkins, J., *B.I.B.R.A. Bull.*, 1979, **18,** pp. 7, 79, 186, 246 and 296.
27. Bartsch, H., Malaveille, C., Barbin, A., Brésil, H., Tomatis, L. and Montesano, R., Mutagenicity and metabolism of vinyl chloride and related compounds, *Environ. Health Perspect.*, 1976, **17,** 193.
28. Stafford, J., The vinyl chloride monomer health problem, *Chemy. Ind.*, 1976, 466.
29. Baxter, P. J., Anthony, P. P., MacSween, R. N. M. and Scheuer, P. J., Angiosarcoma of the liver in Great Britain, 1963–73, *Brit. Med. J.*, 1977, 919.
30. IARC, *Monograph on the evaluation of the carcinogenic risk of chemicals to humans*, Suppl. No. 1, International Agency for Research on Cancer, Lyon, 1979.
31. Haley, T. J., Vinylidene chloride: a review of the literature, *Clin. Toxicol.*, 1975, **8,** 633.
32. Cooper, P., The varying fate of vinylidene chloride, *B.I.B.R.A. Bull.*, 1979, **18,** 522.
33. Anon, Acrylonitrile—focus on hepatic enzymes, *B.I.B.R.A. Bull.*, 1979, **18,** 254.
34. Liebman, K. C., Metabolism and toxicity of styrene, *Environ. Health Perspect.*, 1975, **11,** 115.
35. Cooper, P., Styrene and the chromosome, *B.I.B.R.A. Bull.*, 1980, **19,** 6.

SUGGESTIONS FOR FURTHER READING

Casarett, L. J. and Doull, J. (Eds.), *Toxicology: The basic science of poisons*, Macmillan, New York, Toronto, London, 1975, p. 384.

Committee Report, Proposed system for food safety assessment, *Food Cosmet. Toxicol.*, 1978, **16,** Supplement 2.

Committee Report, *The testing of chemicals for carcinogenicity, mutagenicity and teratogenicity*, Ministry of Health and Welfare, Canada, 1973.

DAVIS, W. and ROSENFELD, C. (Eds.), *Carcinogenic risks, strategics for intervention*, IARC Scientific Publication No. 25, International Agency for Research on Cancer, Lyon, 1979.

GALLI, C. L., PAOLETTI, R. and VETTORAZZI, G., *Chemical toxicology of food*, Developments in Toxicology and Environmental Science, Vol. 3, Elsevier, Amsterdam, 1978.

LEFAUX, R., *Practical toxicology of plastics* (English edn edited by Hopf, P. P.), Iliffe, London, 1968.

MONTESANO, R., BARTSCH, H. and TOMATIS, L., *Screening tests in chemical carcinogenesis*, International Agency for Research on Cancer, Publication No. 12, Lyon, 1976.

Chapter 5

INTERNATIONAL LEGISLATION

INTRODUCTION

The composition and purity of food has been a matter for concern throughout the world for very many years, leading to a highly complex system of legislation and enforcement. In the United Kingdom, attempts to control the adulteration of food commenced in the thirteenth century, and in the early years it was the duty of the Guilds to maintain and protect the standards of various commodities of commerce. In the last hundred years there has been a switch to statutory and scientific methods of control, stimulated both by the progress and developments in analytical chemistry and by an increasing social conscience and awareness on the part of the general public. The primary purpose of food legislation is to ensure that consumers receive food which is: free from adulteration and satisfies minimum standards of composition; free from harmful additives and contaminants; and correctly and informatively labelled and advertised. Further restrictions on permissible practices in factories, slaughterhouses, restaurants, etc., are embodied in the Food Hygiene Regulations and important as these are for public health, they will not be discussed further here since they relate to unpackaged food. Through the enactment and enforcement of such legislation, the authorities aim to protect not only the consumer but also reputable manufacturers from unfair competition. By way of contrast, packaging law is relatively recent in origin.

In the review which follows, legislation pertaining to weights and measures, copyright, Trades Descriptions Act, labelling and permitted sizes and transport of containers have been excluded. Only regulations relevant to the composition of packaging materials will be discussed. Inevitably, a short review of the type that follows cannot claim to be comprehensive, particularly in such a rapidly changing field. Hence, for a detailed and definitive view of the situation in any given country, reference to the original texts is essential.

UNITED KINGDOM

In the UK, the consumer is protected by the general provisions of the Food and Drugs Act, 1955 supported, where necessary, by many detailed regulations prescribing minimum standards of composition for certain foods (e.g., meat products, coffee, cheese, etc.) as well as regulations controlling the use of additives and contaminants (e.g., the Antioxidant in Food Regulations, the Lead in Food Regulations, etc.). No part of the 1955 Act relates specifically to packaging materials. Section 1 of the Act states:

> '1. (1) No person shall add any substance to food, use any substance as an ingredient in the preparation of food, abstract any constituent from food, or subject food to any other process or treatment, so as (in any such case) to render the food INJURIOUS to health with intent that the food shall be sold for human consumption in that state.'

Of particular relevance to this discussion is a later paragraph of the Act which states:

> '(5) In determining for the purposes of this Act whether an article of food is injurious to health, regard shall be had not only to the probable effect of that article on the health of a person consuming it, but also to the probable *cumulative* effect of articles of substantially the same composition on the health of a person consuming such articles in ordinary quantities.'

In recent years, however, there have been very few prosecutions under Section 1 of the 1955 Act and certainly none relating to packaging materials. Obviously, no manufacturer is intentionally going to sell food that is injurious to health. More general protection for the consumer is, therefore, provided under Section 2 of the Act:

> '2. (1) If a person sells to the prejudice of the purchaser any food or drug which is not of the nature, or not of the substance, or not of the quality, of the food or drug demanded by the purchaser, he shall, subject to the provisions of the next following section, be guilty of an offence.'

The phrase—nature, substance or quality—has been enshrined in UK food law for many years and together covers most cases of adulteration, use of improper ingredients or the presence of foreign bodies, etc. Hence, any packaging material that does not adversely affect the flavour or wholesomeness of the food could not be deemed to affect the nature, substance or quality of that food and would, therefore, be permitted. However, the Act gives Ministers powers to make additional regulations as follows:

> '4. The Ministers may, so far as it appears to them to be necessary or expedient in the interests of public health, or otherwise for the protection of the public, make regulations for any of the following purposes:-
>
> (a) for requiring, prohibiting or regulating the addition of any specified substance, or any substance of any specified class, to food intended for sale for human consumption or any class of such food, or the use of any such substance as an ingredient in the preparation of such food, and generally for regulating the composition of such food.'

It is important to note one further paragraph which states:

> '(2) In the exercise of their functions under this section the Ministers shall have regard to the desirability of restricting, so far as is practicable, the use of substances of *no nutritional value* as foods or as ingredients of foods.'

Under UK law an offence takes place only when a food is sold. In Section 1 of the Act it states:

> 'No person shall sell for human consumption, offer, expose or advertise for sale, or have in his possession for the purpose of such sale....'

Hence, packaging manufacturers would not be directly liable under the Act except in so far as they may have given assurances that their product was safe for use with food.

Despite the wide ranging powers contained in the 1955 Act (of which only parts of 3 out of a total of 137 sections have been reproduced above) it was thought that such powers as existed were inappropriate for the making of regulations to control the manufacture and labelling of materials and articles intended to come into contact with food. Such regulations were required to honour UK obligations under the Treaty of Accession into the European Economic Community following the adoption of certain Directives by the Council of Ministers as discussed below. The regulations were, therefore, made under the European Communities Act, 1972. However, for the UK, the principal effect of this new legislation on packaging materials has been to make explicit what is already implicit in the Food and Drug Act, 1955.

In addition to the general, but somewhat vague, provisions of the Food and Drug Act, 1955, a Code of Practice for the Safety in Use of Plastics for Food Contact Applications has been prepared by the British Plastics Federation in co-operation with the British Industrial Biological Research Association. The first edition was issued in 1969 and revisions appeared in 1973 and 1979. This guide relates principally to the use of various ingredients, additives and other processing aids that are used by the manufacturer in the formulation of plastics compositions. The recommendations are based on existing toxicological data and the expected migration behaviour when in use.

Ceramic Articles

Most of the above discussion has been written with plastic packaging materials in mind. However, public concern over the leaching of toxic metals such as lead and cadmium from glazed ceramic ware has resulted in legislation to prohibit the use of those articles, which release excessive amounts of toxic metals. In the UK, the following statutory instruments are in force: S.I. 1972 No. 1957 Cooking Utensils (Safety) Regulations; and S.I. 1975 No. 1241 Glazed Ceramic Ware (Safety) Regulations. These regulations prohibit the use of articles which release up to 20 mg/litre of lead or 2 mg/litre of cadmium in a 24 h leaching test, using 4 per cent acetic acid at 20 °C. A separate hot test is carried out on cooking ware. This aspect is discussed more fully in a later section of this chapter.

EUROPEAN ECONOMIC COMMUNITY

The birth of the European Economic Community, with the signing of the Treaty of Rome in 1957, was an attempt to create suitable conditions for economic growth and political stability, following the legacy of two world wars within the previous thirty years. Initially, much of the effort centred round the removal of tariff barriers between member states, with the objective of promoting easier intra-community trade. Whilst tariff barriers have now been removed, differing technical standards imposed by member states restrict the movement of goods and may create unequal conditions of competition within the Community. Many of these restrictions lie in the field of food manufacture. Hence, the EEC has embarked[1] on a programme of approximation (or harmonisation) of a wide range of technical regulations covering such aspects as the nature, composition, handling, packaging and labelling of foodstuffs. This process is essentially very slow as decisions to implement new harmonised legislation are not taken without the fullest possible consultation amongst the Governments of member states as well as trade associations, etc. The consultative procedures of the Commission, the Council, the Parliament and the Economic and Social Committee have been described in detail by Crosby[2] and, more recently by Haigh.[3] Current progress and future prospects are discussed in a later paper by Haigh.[4] In this field, decisions of the Council are usually promulgated as Directives, which are then binding on member Governments whilst allowing each state to adjust its legislation as it wishes, providing only that the original objectives are fulfilled. Harmonisation of food legislation is usually classified into Vertical Directives (concerned with a specific group of similar products, e.g. coffee extracts, cocoa and chocolate products, sugars, honey or fruit juices) and Horizontal Directives (concerned with subjects of general application to all foods, e.g. additives, contaminants, labelling, packaging, etc.).

However, the main impact of EEC legislation for the packaging manufacturer will arise from the approval by the Council of a Directive on Materials and Articles[5] in November, 1976. This Directive lays down a framework of general principles by means of which legal differences may subsequently be eliminated by specific Directives. Materials that are likely to be the subject of subsequent specific Directives include plastics, ceramics, glass, rubber, regenerated cellulose, etc. Whilst the general Directive applies to materials and articles in contact with water, fixed public or private supply equipment is specifically excluded. Covering or coating substances, e.g. for cheese rinds are also excluded.

Article 2 requires that:

> 'Materials and articles must be manufactured in compliance with good manufacturing practice, so that, under their normal or foreseeable conditions of use, they do not transfer their constituents to foodstuffs in quantities which could:
>
> (1) endanger human health,
> (2) bring about an unacceptable change in the composition of the foodstuffs or a deterioration in the organoleptic characteristics thereof.'

The Directive also provides for labelling of materials and articles, the promulgation and content of specific Directives, and for rules to check compliance with these Directives. Sampling procedures, and methods of analysis required to enforce Directives, are to be established through the Standing Committee for Foodstuffs, using qualified majority voting procedures if necessary.[3] In the UK the general Directive (76/893/EEC) has been implemented by the Materials and Articles in Contact with Food Regulations, 1978 (S.I. 1978 No. 1927) which came into operation on 26 November, 1979.

One specific Directive has so far been adopted by Council.[6] This prohibits the use of materials and articles for food use if they contain more than 1 mg/kg of vinyl chloride in the final product. In addition, the Directive stipulates that:

> 'Materials and articles must not pass on to foodstuffs which are in or have been brought into contact with such materials and articles any vinyl chloride detectable by the method which complies with the criteria laid down in Annex II.'

Annex II specifies the following criteria:

'1. The level of vinyl chloride in materials and articles and the level of vinyl chloride released by materials and articles to foodstuffs are determined by means of gas-phase chromatography using the 'headspace' method.
2. For the purposes of determining vinyl chloride released by materials and articles to foodstuffs, the detection limit shall be 0·01 mg/kg.
3. Vinyl chloride released by materials and articles to foodstuffs is in principle determined in the foodstuffs. When the determination in certain foodstuffs is shown to be impossible for technical reasons, Member States may permit determination by simulants for these particular foodstuffs.'

In the UK, this Directive will be implemented by means of an amendment to the Materials and Articles in Contact with Food Regulations, 1978. This will not be incorporated into UK legislation until agreed methods of analysis have been developed.

The determination of residues of vinyl chloride monomer in plastics and in foodstuffs has already been discussed (Chapter 3). In a method which has to be used for referee action or for legal enforcement, it is essential that the protocol is unambiguous and that the performance characteristics attainable are clearly defined. In particular, for vinyl chloride, problems can arise in the preparation of standard solutions for calibration purposes. Furthermore, the finally agreed procedure must be adequately evaluated by collaborative study using a wide range of samples containing known amounts of vinyl chloride at, or near, the proposed legal limit.

A number of other Directives are currently under discussion and the Commission has made proposals to the Council in respect of ceramics[7] and plastics.[8] Draft proposals have been prepared by the Commission for glass and for regenerated cellulose film. However, only the principal provisions of such proposals will now be described, since they are almost certain to be modified before final acceptance by the Council. (The Community method of analysis for the control of VCM in materials and articles which are intended to come into contact with foodstuffs has now been approved.[9])

Ceramics

The main provision in this proposed Directive is to limit the quantity of lead or cadmium leached from table and kitchenware, cooking ware, and packaging and storage vessels arising from the use of glazing materials. The proposed limits are shown in Table 1. The extractability depends on the method of test adopted. Most tests proposed recommended 4 per cent acetic acid as the extractant, for 24 h at about room temperature. Lead or cadmium is then determined in the extract by atomic absorption spectrophotometry, or colorimetrically using dithizone reagent. Further difficulties arise from the fact that in the case of cadmium it is also necessary to control and specify the conditions of illumination during the test. Experimental aspects of this test will be discussed more fully in a later section. Regulations in other countries throughout the world are summarised in Table 2.

TABLE 1

PROPOSED LIMITS[7] ON THE QUANTITIES OF LEAD AND CADMIUM EXTRACTABLE FROM CERAMIC ARTICLES

(a) Tableware and kitchenware:
flatware Pb 1 ± 0·05 mg/dm², Cd 0·1 ± 0·005 mg/dm²
holloware articles with a capacity up to 5 litres Pb 5 ± 0·25 mg/litre, Cd 0·5 ± 0·025 mg/litre
(b) Plates specially designed for very young children:
Pb 2·5 ± 0·25 mg/litre, Cd 0·25 ± 0·025 mg/litre
(c) Cooking ware:
flatware Pb 0·5 ± 0·025 mg/dm², Cd 0·05 ± 0·002 5 mg/dm²
holloware Pb 2·5 ± 0·25 mg/litre, Cd 0·25 ± 0·025 mg/litre
(d) Packaging and storage vessels:
Pb 2·5 ± 0·25 mg/litre, Cd 0·25 ± 0·025 mg/litre

TABLE 2

METAL RELEASE FROM GLAZED POTTERY

Country	Metals	Acid (per cent acetic)	Tests temperature	Time (h)	Limits and notes	References
Australia	Pb, Cd	4	20 ± 1 °C	24	*Pb* / *Cd* Hollow-ware: 2 ppm / 0·2 ppm Hollow-ware < 1 100 cm³: 7 ppm / 0·7 ppm Flat-ware: 20 ppm / 2·0 ppm Cooking-ware: 7 ppm / 0·7 ppm (Import prohibition)	Statutory Rules 1973; No. 263, Customs Act 1901–1971. 11 Dec. 1973
			boil + cool	2 22		
Austria	Pb	4	room	24	2 mg/litre or 2 mg if vol of container < 1 litre	
Bulgaria	Heavy metals As, Sb, Ba	4	100 °C	½	nil	
Canada	Pb, Cd	4	68 °F	18	7 ppm Pb, 0·5 ppm Cd	Hazardous Products (Glazed Ceramics) Regulations as amended 1971
China (Taiwan)	Pb, Cd	4	—	—	7 ppm Pb, 0·5 ppm Cd	CNS 3725, R164; Test method CNS 3503, R154
Czechoslovakia	Pb, Cd, Zn Sb, As, Cu, Sn	4	room	—	—	CSN 72 5505 and PN201-74 to PN 207-74
Denmark	Pb, Cd	4	boil	3 × ½	3 mg/litre of vol of use Pb, 1 mg/litre Cd 0·6 mg/dm² Pb, 0·2 mg/dm² Cd*	Ministry of Environmental Protection, Statutory Order No. 450, 11 Oct 1972
Eire	Pb, Cd	4	16–23 °C	24	7 μg/ml Pb, 0·5 μg/ml Cd	Irish Institute for Industrial Res. & Standards IS 186, 1973
Finland	Pb, Cd, Sb, Zn,	4	room	24	0·6 mg/dm² total	Government of Finland Foodstuffs, General Regulations
	As				nil	Decree 408 21/11/62, amended by Decree 477, 27/8/65
France	Pb, Cd	4	room	24	7 ppm Pb, 0·5 ppm Cd	Unconfirmed (*Silikat J.* 12, 38, 1973)
Germany, East	Pb	4	20 °C	24	0·4 mg/dm²	Draft TGL 14929, 1963
Germany, West					*Flatware* *Pb* / *Cd* (mg/dm²); *Hollow-ware* *Pb* / *Cd* (mg/litre) Tableware: 1·0 / 0·10; 5·0 / 0·50 Cooking-ware: 0·5 / 0·05; 2·5 / 0·25 Storage vessels: — / —; 2·5 / 0·25 Plus a lip test with limits Pb 2·0 mg, Cd 0·20 mg	*Test method* DIN 51031, 1976 *Limits* DIN 51032, 1977

Holland	Pb	4	room	24	3 mg/dm²	Stattsblatt Koninkrijk der Nederlanden 1949, No. J306
India	Pb	5	boil	½	2 ppm	IS 3505, 1065; IS 2857, 1964
Israel	Pb, Cd	4	22 °C	24 h incl. 8 h in ambient light	*Pb* *Cd* (mg/litre) Deep Tableware (> 25 mm interior height) 3 0·3 Shallow Tableware 5 0·5 Children's Tableware 0·4 0·04 Storage Containers (incl. vessels with vol > 5 litre) 0·4 0·04	
		4	boil (warm vessel to 120 °C first)	2·5 + 22 h ambient	Cooking-ware 3 0·3 (The above classes to be marked on the ware)	
Italy	Pb	1	room	24	0 ('temporary' (1964) working limit 0·5 mg/dm²)	Law of April 30, 1962 no. 283 Gazetta Ufficiale della Republica Italiano 1962, Pt. 1. (139) p. 2194
Japan	Pb, Cd				*Pb* *Cd* / *Pb* *Cd* Flat-ware 20 ppm 0·5 ppm; Hollow-ware 7 ppm 0·5 ppm Containers < 1 l 7 ppm 0·5 ppm; Containers > 1 l 2 ppm 0·5 ppm	Japan Ceramic Industry proposal, not law (*Ber. Dtsch. Keram. Ges.* Feb. 1974)
New Zealand	Pb, Cd, Sb, As	4	room	24	h/d ratio < ½: 20 ppm Pb, ¨ppm Cd, Sb, As h/d ≥ ½: usable vol ≤ 1 l, 7 ppm Pb, 0·7 ppm Cd, Sb, As; usable vol > 1 l, 2 ppm Pb, 0·2 ppm Cd, Sb, As	N.Z. Food and Drug Regs, 1973, Reg. 43
		4	boil cool	2 22	Cooking-ware 7 ppm Pb, 0·7 ppm Cd, nil Sb, As	
Norway		4	room	24	Hollow-ware 2 ppm Pb, 0·1 ppm Cd; flat-ware 1 mg/dm² Pb, 0·05 mg/dm² Cd†	
Poland	Pb	4 10 (citric)	room 100 °C	24 4	0·15 mg/dm²	PN-66 C12490, test method BN-65/7001-02
Portugal	Pb	4	60 °C and cool	24	7 mg/dm³ of solution Pb, 0·5 mg/dm³ of sol. Cd	Limits and tests, Portuguese draft standards 1-1180–1-1184, 1972
Rumania		4	80 °C	½		STAS 708-49 (test method only)
South Africa	see 'limits'	4	boil	½	1 Sb, 1 As, 1 Cd, 20 Cu, 1 F, 1 Pb, 50 Ni, 250 Sn, 50 Zn mg/kg	Believed obsolete. BS 4860 used
South Korea	Pb, Cd	4	90 °C room	½ 24	*Pb* *Cd* Cooking-ware 60 °C Steaming 2·0 ppm Nil Tableware	Does not dist. between pottery, glass and enamels, Min. Reg. Section 458 13/9/74
Sweden	Pb, Cd	4 4	boil room	3 × ½ 24	3 mg/litre of vessel capacity, of Pb 0·1 mg Cd. Total ban on import and use of Cd as pigment from July 1, 1980	

TABLE 2—*contd.*

Country	Metals	Acid (per cent acetic)	Tests temperature	Time (h)	Limits and notes	References
Switzerland	Pb, Cd, Zn	4	room	24	3 mg/dm² Pb + Cd + Zn inside + 2 cm lip 100 mg/100 cm² total surface	Swiss Food Ordinance Part 453
Thailand	Pb	4	boil	½	2 ppm	TIS 32—1973 (standard)
UK					*Pb* *Cd*	SI 1241, 1975 Consumer Protection. Glazed Ceramic Ware Safety Regulations 1975 & BS 4860 Parts 1 and 2, 1972
	Pb	4	room	24	Flat-ware 20 mg/litre 2 mg/litre	
		4	boil 100 °C		Hollow-ware < 1·1 litre 7 mg/litre 0·7 mg/litre	
	Cd		+ cool		≥ 1·1 litre 2 mg/litre 0·2 mg/litre	
		Test for Cd to exclude light			Cooking-ware 7 mg/litre 0·7 mg/litre	
USA (FDA)	Pb, Cd	4	room	24	*Pb* *Cd*	Food and Drug Administration. Food Drug and Cosmetic Act Sections 402(a) (2)
					Flatware 7·0 ppm 0·5 ppm	
					Small hollow-ware (< 1·1 litre) 5·0 ppm 0·5 ppm	
					Large hollow-ware (> 1·1 litre) 2·5 ppm 0·25 ppm	
USA (California)	Pb, Cd	4	25 ± 2 °C	24	7 ppm Pb, 0·5 ppm Cd	
USA (New York, Monroe County)	Pb, Cd	4	68 °F	24	7 ppm Pb, 0·5 ppm Cd	Monroe County Sanitary Code
Yugoslavia	see 'limits'	4	boil	½	Vol > 1 l, 4·0 mg/l Pb; vol 100 ml – 1 l, 0·5 mg/l	
			or room	24	Also colours, on chemical analysis, must contain, in total per 100 g dry colour or per 1 m² of decorated surface: < 0·05 g As and < 0·2 g Sb, Ba, Pb, Cr, Cd, Cu, Zn, CaSO, talij, cyanogen alloys, oxalic acid	
		4	room	24	5 mg/dm² Pb*	

* Lip test † Two tests are now proposed: (1) 4% acetic at 20 ± 2 °C for 24 h.
(2) 4% acetic, 1-h boil, left to cool in room at 20 ± 2 °C for 24 h.
In both (1) and (2) if Cd is tested, samples must be illuminated for 1-h with 1 000 + 10% lux, then light excluded.

Plastics

The proposed Directive is applicable to all types of plastics, including laminates, where the plastic is in direct contact with the foodstuff. Regenerated cellulose film, elastomers, rubbers, surface coatings (e.g. varnishes, lacquers, etc.), adhesives, paper and paperboard impregnated with plastics materials are all excluded. Initially, the Directive proposes to limit the overall migration of constituents of plastics into foodstuffs to a value of 60 ppm. For liquids, the limit would be 60 mg/litre, and for materials and articles the limit is expressed as 10 mg/dm^2. The Directive also lays down basic rules for the determination of the overall migration by using simulants. Test conditions (times and temperatures of contact) are also specified, as shown in Table 3. The theoretical aspects and experimental determination of overall migration will be fully discussed later.

Control of plastics materials is likely to be extended by the adoption of further Directives covering (a) a positive list of ingredients (especially monomers), with associated purity criteria, used in the manufacture of plastics, and (b) the classification of foodstuffs. The second provision is required to cover a large number of foodstuffs which are mixtures of aqueous and fatty systems. The classification would identify not only the appropriate simulant (or simulants) to be used in testing, but would also define an agreed factor to be applied to the results obtained in any such test.

Four separate simulating solvents have been proposed, *viz:*

A—distilled water for aqueous foods
B—3 per cent (w/v) acetic acid for acidic foods, e.g. vinegar, pickles
C—olive oil for fatty foods
D—15 per cent (v/v) ethanol for spirits

These simulants are designed to represent the extreme conditions of extraction encountered with the types of food listed. However, most foods cover a whole spectrum in composition, of which the above categories are the extremities. Hence, a classification of foodstuffs is required so that a weighting coefficient can be assigned to each individual food (or group of foods) to determine how nearly it approaches the extremity of its class, with respect to extraction capability using a particular simulating solvent. Thus, a fermented full-fat milk product such as yoghurt would need to be tested with both aqueous and fatty simulants. However, the result (M) obtained with the fatty simulant would be subject to a weighting coefficient (say $M \div 3$) to allow for the greater extractive capacity of the simulant

TABLE 3

TEST CONDITIONS TO BE CHOSEN IN KEEPING WITH THE PRACTICAL CONDITIONS OF CONTACT TIME (t) AND TEMPERATURE (T) IN ACTUAL USE

Conditions of contact in practice	Test conditions	Materials and articles
1. Prolonged contact ($t > 24$ h)		
1.1. at low temperature ($T < 5$ °C)	10 days at 5 °C	e.g. containers, accessories for containers and closing devices (bottles, jars, pots, tubes, tins, drums, demi-johns, boxes, baskets and receptacles in general; valves, rings, safety plating, linings, seals, etc.; stoppers, lids, capsules, etc.); packaging film
1.2. at medium temperature (5 °C $\leq T \leq$ 40 °C)	10 days at 40 °C	e.g. see 1.1.
2. Long contact (2 h $\leq t \leq$ 24 h)		
2.1. at low temperature ($T <$ 5 °C)	24 h at 5 °C	
2.2. at medium temperature (5 °C $\leq T \leq$ 40 °C)	24 h at 40 °C	e.g. kitchen crockery (plates, cups, glasses, etc.)
3. Brief contact ($t <$ 2 h)		
3.1. at low temperature ($T <$ 5 °C)	2 h at 5 °C	e.g. tubes, piping, conveyor belts, funnels, siphons and articles in general which are in dynamic contact with the foodstuff; kitchen utensils
3.2. at medium temperature (5 °C $\leq T \leq$ 40 °C)	2 h at 40 °C	e.g. see 3.1.
3.3. at high temperature (40 °C $< T \leq$ 70 °C)	2 h at 70 °C	e.g. articles in contact with foodstuffs that are undergoing pasteurisation, others
3.4. at high temperature (70 °C $< T \leq$ 100 °C)	1 h at 100 °C	e.g. kitchen utensils (saucepans, frying pans and other articles for cooking and frying foods)
3.5. at high temperature ($T >$ 100 °C)	30 min at 121 °C[a,b]	e.g. articles in contact with foodstuffs that are undergoing sterilisation

[a] Where the heating of the foodstuff is preceded or followed by a period of prolonged contact, the test must be preceded or followed by the corresponding transference test (10 days). In such a case, the same simulant should be used if possible.
[b] Only the simulants A, B and D are to be used for this test.

compared to the yoghurt itself. A similar situation arises with foodstuffs that are not in continuous contact with the plastic material or article but touch the container only in places.

UNITED STATES OF AMERICA

The corresponding law in the USA is the Food, Drugs and Cosmetics Act, 1938. American legal requirements are published in *The Code of Federal Regulations*.

The Code is divided into 50 titles, or broad areas, subject to federal control. Title 21 is concerned with food and drugs and is made up of seven volumes. The most important provisions relating to packaging materials can be found in Volume II, Parts 100–199. For example, in Subpart E §110.80(h) it states:

> 'Packaging processes and materials shall not transmit contaminants or objectionable substances to the products, shall conform to any applicable food additive regulation (Parts 170–189) and should provide adequate protection from contamination.'

The definition of a food additive can be found in §170.3. Included in this paragraph is the statement: 'A material used in the production of containers

TABLE 4
TYPES OF FOOD
(Code of Federal Regulations, 21 CFR§175.300)

I	Non-acidic (pH $>5{\cdot}0$), aqueous products; may contain salt or sugar or both, and including oil-in-water emulsions of low- or high-fat content.
II	Acidic (pH $\leq 5{\cdot}0$), aqueous products; may contain salt or sugar or both, and including oil-in-water emulsions of low- or high-fat content.
III	Aqueous, acidic or non-acidic products containing free oil or fat; may contain salt, and including water-in-oil emulsions of low- or high-fat content.
IV	Dairy products and modifications:
	(a) Water-in-oil emulsion, high- or low-fat.
	(b) Oil-in-water emulsion, high- or low-fat.
V	Low moisture fats and oils.
VI	Beverages:
	(a) Containing alcohol.
	(b) Non-alcoholic.
VII	Bakery products.
VIII	Dry solids (no end test required).

TABLE 5

TEST PROCEDURES FOR DETERMINING THE AMOUNT OF EXTRACTIVES FROM RESINOUS OR POLYMERIC COATINGS, USING SOLVENTS SIMULATING TYPES OF FOODS AND BEVERAGES (Code of Federal Regulations, 21 CFR §175.300)

Condition of use	Types of food[a]	Extractant		
		Water (temperature and time)	Heptane (temperature and time)	8 per cent alcohol (temperature and time
A. High temperature heat-sterilised	I, IV(b)	250 °F (120 °C), 2 h	—	—
(e.g. 212 °F (100 °C))	III, IV(a), VII	250 °F (120 °C), 2 h	150 °F (66 °C), 2 h	—
B. Boiling water sterilised	II	212 °F (100 °C), 30 min	—	—
	III, VII	212 °F (100 °C), 30 min	120 °F (49 °C), 30 min	—
C. Hot filled or pasteurised above 150 °F (66 °C)	II, IV(b)	Fill boiling, cool to 100 °F (38 °C)	—	—
	III, IV(a)	Fill boiling, cool to 100 °F (38 °C)	120 °F (49 °C), 15 min	—
	V	—	120 °F (49 °C), 15 min	—
D. Hot filled or pasteurised below 150 °F (66 °C)	II, IV(b), VI(b)	150 °F (66 °C), 2 h	—	—
	III, IV(a)	150 °F (66 °C), 2 h	100 °F (38 °C), 30 min	—
	V	—	100 °F (38 °C), 30 min	150 °F (66 °C), 2 h
	VI(a)	—	—	—
E. Room temp filled and stored (no thermal treatment in the container)	I, II, IV(b), VI(b)	120 °F (49 °C), 24 h	—	—
	III, IV(a)	120 °F (49 °C), 24 h	70 °F (21 °C), 30 min	—
	V, VII	—	70 °F (21 °C), 30 min	120 °F (49 °C), 24 h
	VI(a)	—	—	—
F. Refrigerated storage (no thermal treatment in the container)	I, II, III, IV(a)	70 °F (21 °C), 48 h	—	—
	IV(b), VI(b)	—	—	—
	VII	—	—	—
	VI(a)	—	—	70 °F (21 °C), 48 h
G. Frozen storage (no thermal treatment in the container)	I, II, III, IV(b)	70 °F (21 °C), 24 h	—	—
	VII			
H. Frozen storage: Ready-prepared foods intended to be reheated in container at time of use:				
1. Aqueous or oil in water emulsion	I, II, IV(b)	212 °F (100 °C), 30 min	—	—
2. Aqueous high or low free oil or fat	III, IV(a), VII	212 °F (100 °C), 30 min	120 °F (49 °C), 30 min	—

and packages is subject to the definition if it may reasonably be expected to become a component, or to effect the characteristics, directly or indirectly, of food packed in the container.' Also, 'If there is no migration of a packaging component from the package to the food, it does not become a component of the food and thus is not a food additive.'

In 1958, Congress passed the Food Additive Amendment of the 1938 Act. This amendment required the manufacturer to establish the safety of any product about to be marketed and the Government had the responsibility to check the evidence of safety supplied. Section 409(c)(3)(A) of the amendment often referred to generally as the Delaney clause, states:

> 'That no additive shall be deemed to be safe if it is found to induce cancer when ingested by man or animal, or if it is found, after tests which are appropriate for the evaluation of the safety of food additives, to induce cancer in man or animal....'

However, at the same time Congress provided for the approval of commonly used food ingredients by defining Generally Recognised as Safe (GRAS) substances as well as Prior-Sanctioned Food Ingredients (i.e., those approved before 6 September, 1958). The latter category contains certain substances employed in the manufacture of food packaging materials (§181.22). Such substances are excluded from the definition of a food additive provided that they are of good commercial grade, are suitable for association with food, and are used in accordance with good manufacturing practice. The Code includes lists of specific antioxidants, plasticisers, release agents, stabilisers, etc., as well as copolymers and resins. Limits for acrylonitrile monomer extraction are specified; determined by using simulating solvents appropriate to the conditions of use.

Selection of extractability conditions depend (a) on the type of food, and (b) on the conditions of use and, in particular, the thermal treatment used at the packaging stage. The classification of food is described in Table 4 and appropriate test procedures for different conditions of use in Table 5.

OTHER COUNTRIES

In general, the Commonwealth countries have followed UK legislation, whilst other countries have adopted simplified versions of USA law. Whilst some European countries, e.g. Belgium, Denmark, France, Holland, Italy and West Germany have their own special regulations relating to food packaging, inevitably as members of the nine EEC countries they will adopt similar standards following the approval of the EEC Directive in the near future. With Greece, Portugal and Spain as future members of the Community, this will impose a degree of uniformity throughout the European continent which should assist in the removal of trade barriers.

Currently, control of packaging materials throughout the world is exercised by two principal systems:

(1) A legally enforceable positive list of permitted and approved products with well defined extractability tests, or
(2) Reliance on Codes of Practice as in the UK and West Germany.

There are also differences in enforcement practices. In the UK and the Netherlands, there is in existence an extensive system of food inspectors with access to a wide range of laboratory facilities. In other countries, more reliance is placed on enforcement by approval of one central authority acting on administrative and legal considerations, rather than practical scientific evidence.

Perhaps the area of greatest concern at the present time, whatever the control system adopted, is the responsibility for guarantees as to compliance with the legal requirements. Food manufacturers and packers will naturally seek assurances from the suppliers of packaging materials that a particular product complies with the relevant regulations and is safe to use in contact with food. Such suppliers, polymer manufacturers or converters, in turn may not be able to give such guarantees since they have no control over subsequent processing or end-use. Hence, there is a danger of the buck passing from end to end, along the line. In this situation there is a lot to be said for Codes of Practice administered by a single competent authority in co-operation with the industry and with government, with powers to arbitrate in doubtful cases.

Since the transfer process from the plastic into the foodstuff is such a crucial factor in the discussion as to whether a hazard exists or not, both the theoretical aspects as well as the experimental determination of migration will now be considered in some detail.

REFERENCES

1. Resolution of the Council of the EEC of 17 December 1973 on Industrial Policy. *Official Journal of the EEC*, No. C117, 31 December, 1973.
2. Crosby, N. T., Entry into the EEC and its effect on official methods of analysis, *J. Assoc. Publ. Analysts*, 1973, **11**, 92.

3. HAIGH, R., Harmonization of legislation on foodstuffs, food additives and contaminants in the EEC, *J. Food Technol.*, 1978, **13,** 255.
4. HAIGH, R., Achievements and programme, *J. Food Technol.*, 1978, **13,** 491.
5. Council Directive of 23 November 1976 on the approximation of the laws of the Member States relating to Materials and Articles intended to come into contact with foodstuffs, *Official Journal of the EEC*, No. L340, 9 December, 1976.
6. Council Directive of 30 January 1978 on the approximation of the laws of the Member States relating to Materials and Articles which contain vinyl chloride monomer and are intended to come into contact with foodstuffs, *Official Journal of the EEC*, No. L44, 15 February, 1978.
7. Proposal for a Council Directive on the approximation of the laws of the Member States relating to ceramic articles intended to come into contact with food (limitation of extractable quantities of lead and cadmium), *Official Journal of the EEC*, No. C46, 27 February, 1975.
8. Proposal for a specific Council Directive on the overall migration limit for the constituents of plastics materials and articles intended to come into contact with foodstuffs, *Official Journal of the EEC*, No. C141, 16 June, 1978.
9. Commission Directive of 8 July, 1980 laying down the Community method of analysis for the official control of the vinyl chloride monomer level in materials and articles which are intended to come into contact with foodstuffs, *Official Journal of the EEC*, No. L213/42, 16 August, 1980.

SUGGESTIONS FOR FURTHER READING

PAINE, F. A. (Ed.), *Packaging and the law*, Newnes-Butterworths, London, 1973.
O'KEEFE, J. A., *Sale of food and drugs*, Fourteenth edn, Butterworths, London, 1968.

Chapter 6

MIGRATION—THEORETICAL ASPECTS

INTRODUCTION

Some definitions and physico-chemical concepts and laws commonly encountered in the literature on migration will first be discussed. Mathematical and pictorial models of migration will then be considered along with the experimental evidence that has been accumulated in support of each different approach.

DEFINITIONS

In food packaging terminology, *migration* is used to describe the transfer of substances from the package (usually a plastic material) into the foodstuff. Substances that are transferred to the food as a result of contact or interaction between the food and the packaging material are often referred to as *migrants* and usually consist of monomers (or residual reactants) and processing additives. Migration is a two-way process, since constituents of the food can also penetrate into the plastic. A distinction is usually made between *global* migration and *specific* migration. Global (or total) migration is the sum of all (usually unknown) mobile packaging components transferred into the food, whereas specific migration relates to one or two individual and identifiable compounds only, either with a particular toxicological interest, or as labelled compounds used in experiments designed to elucidate the extent or mechanism of migration. Global migration, therefore, is a measure of all compounds transferred into the food whether they are of toxicological interest or not, and will include substances that are physiologically harmless. Whilst restrictions on substances of no toxicological importance may seem unnecessary, in the UK food regulations at least, there is a specific requirement that Ministers should have regard to the desirability of restricting, so far as is practicable, the use of substances of no nutritional value as food, or as ingredients of foods (Chapter 5). Equally such substances, even if toxicologically harmless, might have an effect on the organoleptic characteristics of the food. Furthermore, the concept of global migration obviates the need to identify each and every individual migrant, particularly since a complex matrix such as food presents such obvious difficulties to the analytical chemist. Migration testing is a long, laborious, skilled and expensive laboratory operation. The global migration test can be seen as a screening test which saves time and expensive analytical resources, where no particular hazard is expected.

The transference of organic molecules from the packaging material into the food is undoubtedly a complex phenomenon. Most mathematical treatments of transport processes are derived initially from a consideration of gaseous diffusion, where the molecules perform random motions following the concepts of the kinetic theory of gases. Thus, diffusion is a spreading out and can be observed to take place against both gravity and through porous barriers. In liquids, the cohesional forces are much stronger and the molecules are closer together than for gases so that approximately 95 per cent of the space in a liquid is occupied by the molecules themselves. There is still considerable freedom for motion, although diffusion coefficients (see later) in liquids are of the order of $10^{-5}\,cm^2/s$ only, i.e. approximately one million times slower than in gases. In solids, most molecules are fixed in a crystalline regular lattice, except at the surface where imperfections may occur. Diffusion coefficients in solids are much lower still—around $10^{-11}\,cm^2/s$.

Diffusion of gaseous components, e.g. water vapour, air, oxygen, chemical taints from the environment, into foodstuffs, will not be discussed further here, although they too can affect the organoleptic qualities of the food. Many of the same mathematical treatments and theoretical models are equally applicable to this problem area and to the drying of foods. However, the argument will be restricted, for simplicity, to the transfer of components from the packaging material into the food. In the text that follows, some physical–chemical laws that are essential for an understanding of the migration process will first be explained. The application of such theoretical concepts to experimental data will then be discussed and an attempt will be made to relate the results and conclusions obtained to the practical food packaging situation.

SOME PHYSICO-CHEMICAL PRINCIPLES

Molecular Theory of Diffusion

Whilst the transfer of substances from plastic packaging materials into foodstuffs is undoubtedly a complex process, diffusion is thought to be the main controlling mechanism. Diffusion is the mass transfer resulting from the spontaneous natural molecular movements that occur without the assistance of external forces such as shaking, mixing or even convection currents in liquids. It is thus a homogenisation, or an approach to equilibrium, brought about by random atomic or molecular motion. Diffusion processes can be demonstrated experimentally by carefully placing two miscible liquids (one coloured) in contact with each other so that the minimum mixing occurs. After a time, one can see that the coloured liquid has slowly spread into the colourless component and, at equilibrium, will be evenly dispersed throughout the whole vessel. This process occurs most rapidly in gases, only slowly in liquids and infinitesimally slowly in the case of two solids in contact. There is thus a natural tendency for materials to diffuse from areas of high concentration to areas of low concentration until at equilibrium there will be no concentration gradient. More correctly it is usual to state that diffusion takes place from areas of high chemical potential into areas of low chemical potential until an overall constant chemical potential is attained; the resulting ratio of concentrations in the two phases being determined by the Nernst partition coefficient. In an ideal mixture:

$$\mu_i = \mu_i^0 + RT \ln x_i$$

where, μ_i = chemical potential of component i, μ_i^0 = standard chemical potential of i, R = molecular gas constant, T = temperature and x_i = mole fraction of i.

Most real systems do not exhibit ideal behaviour so that

$$\mu = \mu_i^0 + RT \ln a_i$$

where, a_i = activity of component i, i.e., the concentration of i in the ideal mixture which possesses the same chemical potential as in the real mixture.

The relationship between a_i and x_i is expressed in the form:

$$a_i = x_i f$$

where, f = activity coefficient and is a measure of the deviation from ideal behaviour.

In an analogous fashion to changes in concentration, where differences in temperature exist across a boundary, heat will flow from a point of higher temperature to one of lower temperature. Similar changes take place where pressure differentials exist. Chemical potential is, therefore, a measure of the tendency of a component to migrate through the concentration gradient.

Migration is a two-way process which continues until the chemical potentials in the two phases are equal. Not only can components in the plastic diffuse into the foodstuff, it may also be possible for the food to penetrate into the plastic and this can have a great influence on the diffusion rate. This reverse process will depend on the random motion of the plastic polymer chains which may create momentary voids, into which the liquid food can penetrate. The food may also behave as a plasticiser when in contact with the plastic, thus permitting increased deformation of the molecular chains.

Fick's Laws of Diffusion

The rate of the diffusion process is related to the concentration gradient between the two phases in contact. This is known as Fick's first law, and it can be expressed mathematically as follows:

$$\frac{1}{A}\frac{\delta m}{\delta t} = -K_1\frac{\delta c}{\delta t} \qquad (1)$$

where, m = mass of component transferred, t = time, c = concentration, K_1 = diffusion constant, often given the symbol D and A = area of plane, across which diffusion occurs.

The negative diffusion constant arises from the fact that the concentration falls as mass transfer occurs. The process can be illustrated diagrammatically as follows, leading to Fick's second law.

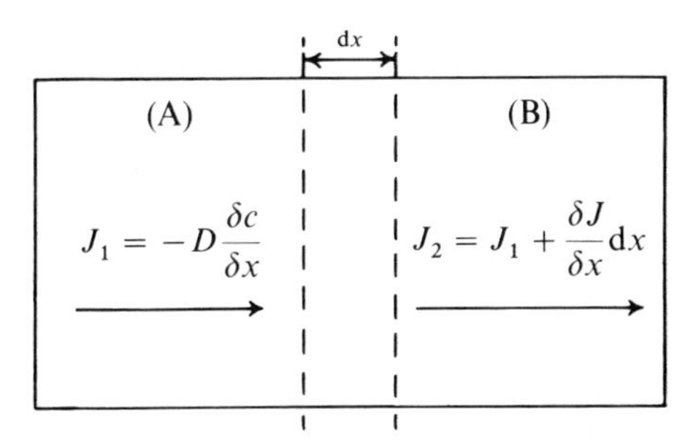

Diffusion across a plane

Let J = flux (quantity/unit time/unit area) perpendicular to the concentration gradient, x = direction of diffusion, c = concentration/unit volume and D = diffusion coefficient.

Therefore, the flux through the first plane (A) is:

$$J_1 = -D\frac{\delta c}{\delta x}$$

and the flux through the second plane (B) is:

$$J_2 = J_1 + \frac{\delta J}{\delta x}\mathrm{d}x$$

$$= -D\frac{\delta c}{\delta x} - \frac{\delta}{\delta x}\left(D\frac{\delta c}{\delta x}\right)\mathrm{d}x$$

subtracting, $J_1 - J_2$:

$$\frac{\delta J}{\delta x} = -\frac{\delta}{\delta x}\left(D\frac{\delta c}{\delta x}\right)$$

Change in flux with distance $= -(\delta c/\delta t)$, therefore:

$$\frac{\delta c}{\delta t} = \frac{\delta}{\delta x}\left(D\frac{\delta c}{\delta x}\right)$$

If D is constant and independent of c, this equation can be written as

$$\frac{\delta c}{\delta t} = D\frac{\delta^2 c}{\delta x^2} \qquad (2)$$

This is known as Fick's second law.

Equation (2) can also be derived from eqn (1) as follows: $C = m/v = m/Ax$ for the unit layer across which diffusion takes place. Therefore:

$$\frac{\delta c}{\delta t} = \frac{-1}{A}\frac{\delta}{\delta x}\left(\frac{\delta m}{\delta t}\right)$$

$$= -\frac{\delta}{\delta x}\left(\frac{1}{A}\frac{\delta m}{\delta t}\right)$$

And from eqn (1):

$$\frac{\delta c}{\delta t} = -\frac{\delta}{\delta x}\left(-K_1\frac{\delta c}{\delta x}\right) = -K_1\frac{\delta^2 c}{\delta x^2}$$

Hence, for steady state diffusion with a fixed concentration gradient (no swelling of polymer by the food) solutions of Fick's second law lead to a determination of the concentration as a function of position and time.

Nernst–Einstein Equation

Since the force causing diffusion is the negative gradient of the chemical potential (μ), rather than the concentration gradient, and if the velocity (v) obtained under the action of a unit force is B; the force of the chemical potential gradient produces a drift velocity and resultant flux:

$$-B = \frac{\text{velocity}}{\text{force}} = \frac{v}{\frac{1}{N}\frac{\mathrm{d}\mu}{\mathrm{d}x}} \tag{3}$$

where N = Avogadro's number and

$$J = -\frac{1}{N}\frac{\mathrm{d}\mu}{\mathrm{d}x}Bc$$

Assuming a unit activity coefficient, the change in chemical potential is given by:

$$\mathrm{d}\mu = RT\,\mathrm{d}\ln c$$

Substituting in eqn (3) and comparing with $J = -D(\delta c/\delta x)$, the diffusion coefficient is directly proportional to the chemical potential.

$$J = -\frac{RT\,\mathrm{d}c}{N\,\mathrm{d}x}B$$

and

$$D = kTB$$

where k = Boltzmann's constant.

The Nernst–Einstein equation has been applied to problems concerning the mobility of charged particles and to the relationship between the diffusion coefficient and electrical conductivity. There are similarities between Fick's laws and Ohm's law, which states that electrical current is proportional to the gradient of electrical potential and to Fourier's law, in which the rate of heat flow is found to be proportional to the temperature gradient.

Solutions of Fick's Laws

Garlanda and Masoero[1] applied the diffusion theory to the migration of constituents from plastics into solvents, to determine the effect on migration of parameters such as time, temperature, polymer thickness, etc. They considered the case of a polymer sheet containing a constituent migrating along the x-axis, perpendicular to the surface of the sheet. At the commencement of the experiment the concentrations of additive within, and at the surface of, the sheet (thickness h) are C_0 and $c = 0$, respectively. Assuming that the sheet is then immersed in a relatively large volume of extractant,

$$\Delta S = \frac{2C_0}{\sqrt{\pi}} \bar{D}t$$

where ΔS = quantity migrating in time t.

Hence, the quantity migrating is proportional to the initial concentration and to $\sqrt{t}$, and is independent of thickness in the initial stages of the experiment. Obviously, this relationship holds for shorter time periods in the case of thin films and longer periods for thick films.

Garlanda and Masoero[1] also considered the general solution of Fick's equations assuming D is constant, i.e. no penetration of the polymer by the food. This treatment gives an average concentration of a constituent $\bar{C}$ as:

$$\bar{C} = \frac{1}{h}\int_0^h C\,\mathrm{d}x = \frac{8C_0}{\pi^2}\sum_{n=0}^{\infty}\frac{1}{(2n+1)^2}e^{-(2n+1)^2\pi^2}h^{-2}Dt$$

Thus, after the initial stages, the linear equation with respect to $\sqrt{t}$ is replaced by a line curving rapidly towards the x-axis until it reaches an equilibrium value. This is the shape of the curve observed in practice. Assuming a coefficient of diffusion (D) equal to $10^{-12}\,\mathrm{cm}^2/\mathrm{s}$ and films of thickness 10, 50 and 100 μm, it was possible to relate migration with time for different thicknesses of polymer and to compute different equilibrium levels. Thus, the equilibrium value of migration is a function of the initial concentration in the polymer. The time taken to reach this limiting value varies with the coefficient of diffusion and the thickness of the sample.

Davies[2] investigated the migration of styrene monomer into foods such as orange squash, margarine and single and double cream. Starting with Fick's diffusion equation, he adapted a solution of the equation derived by Crank,[3] for the situation in which a solid sheet of polymer is in contact with a stirred solution. He obtained an equation of the form:

$$M_t = M_\infty(1 - a\exp(bt))$$

where M_t = amount of styrene which has migrated at time t, and M_∞ = amount of styrene which has migrated at equilibrium. This cannot be determined experimentally since cream deteriorates rapidly even at 5 °C. a and b are constants, and exp is the exponential function.

The theoretical treatment was evaluated by using a curve fitting technique, to determine how well the experimental data coincided with a series of calculated curves, in which unknown constants were varied systematically. In the case of aqueous systems the fit was very good. For fatty products, however, no fit was obtained since the diffusion coefficients were increased by the penetration of the fatty component into the polymer.

Robinson and Becker[4] calculated the diffusion coefficient (D) of an oil entering a polyethylene film using the equation:

$$\ln \frac{(Cs - c)}{Cs} = \ln \frac{8}{\pi^2} - \frac{\pi^2}{h^2} Dt$$

also derived from Fick's laws where, Cs = equilibrium concentration of oil in plastic, c = concentration of oil in plastic at time t and h = thickness of the plastic. The value of D was found to be $26 \times 10^{-10}\,\text{cm}^2/\text{s}$ at 30 °C.

Effect of Temperature

The Arrhenius equation is used in physical chemistry to describe the effect of temperature on the rate of a chemical reaction. The equation has the form:

$$k = A \exp -(E/RT)$$

where k = the rate of reaction, A = frequency factor, E = energy of activation of the reaction, R = gas constant and T = absolute temperature.

Differentiation of the above equation leads to the expression:

$$\frac{\mathrm{d} \ln k}{\mathrm{d}T} = \frac{1}{k}\frac{\mathrm{d}k}{\mathrm{d}T} = \frac{E}{RT^2}$$

which is closely related to the van't Hoff equation derived from thermodynamic considerations to express the effect of temperature on the equilibrium constant of a reversible reaction, and to the Clausius–Clapeyron equation relating vapour pressure and temperature.

In terms of diffusion, the diffusion coefficient is related to temperature by an expression of the type:

$$D = D_0 \exp -(E_D/RT)$$

where superscript E_D = energy of diffusion, which will vary from one system to another. Thus, the quantity diffusing into two solvents S_1 and S_2 at a given temperature can be determined experimentally. The ratio $S_1 : S_2$

may not be the same at a different temperature. This factor makes it very difficult to make comparisons between work reported in the literature at different temperatures and strengthens the case for having fixed conditions of temperature and time for extractability tests (see later).

Diffusion in Systems where the Extractant Penetrates into the Polymer

Where the extractant (or food) penetrates into the polymer, a boundary layer is created. The boundary phase is, in effect, swollen polymer and the diffusion coefficient in this phase will not be the same as that for the unswollen polymer. The swollen layer is inhomogeneous and normal Fickian diffusion equations do not apply since the phase boundary (and hence the diffusion coefficient and chemical potential) are changing with time. Unfortunately, the mathematical expressions derived for such systems are excessively complicated. Frequently, they contain so many constants that are difficult to determine experimentally that the equations are of little practical use. Some simplification is possible by consideration of special cases in which certain assumptions have to be made. Such a treatment has been described by Rudolph[5] and applied to experimental data by Figge and Rudolph.[6] They described the situation in which a plastics additive (C) diffuses from the plastic (B) into a food (A) through a swollen layer produced by A interacting with B. The following equations:

$$C_C^A(\eta) = (C_C^B/\Delta)\, erfc(-\eta/\sqrt{D_C^A}) \qquad \text{for } -\infty < \eta < 0$$

$$C_C^{A+B}(\eta) = (C_C^B/\Delta)[(2\sqrt{D_C^A}/\sqrt{\pi})\,\mathrm{Int}(\eta) + k_1] \qquad \text{for } 0 < \eta < \eta_x$$

$$C_C^B(\eta) = C_C^B\left\{1 - \left[1 - \frac{k_2}{\Delta}\left(\frac{2\sqrt{D_C^A}}{\sqrt{\pi}}\mathrm{Int}(\eta_x) + k_1\right)\right]\frac{erfc(\eta/\sqrt{D_C^B})}{erfc(\eta_x/\sqrt{D_C^B})}\right\}$$
$$\text{for } \eta_x < \eta < \infty$$

in which:

$$\Delta = \frac{\sqrt{D_C^A}}{\sqrt{D_C^B}}\exp\left(\frac{\eta_x^2}{D_C^B} - F(\eta_x^2)\right) erfc\left(\frac{\eta_x}{\sqrt{D_C^B}}\right) + \left(\frac{2\sqrt{D_C^A}}{\sqrt{\pi}}\mathrm{Int}(\eta_x) + k_1\right)$$
$$\times\left[k_2 - \frac{\sqrt{\pi}}{\sqrt{D_C^B}}\eta_x(k_2 - 1)\exp\left(\frac{\eta_x^2}{D_C^B}\right) erfc\frac{\eta_x}{\sqrt{D_C^B}}\right]$$

enable one to calculate the concentration of C at any point x for any given time interval t (see Rudolph[5] for an explanation of the other symbols used). However, it is essential to check that in any normal plastic/food contact situation that such equations, and the assumptions made, do in fact hold and that the migration profiles described by the equations are realistic in practical terms. This treatment emphasises the complexity of the situation where polymer swelling occurs.

Nernst Distribution Law

The diffusion coefficient is a measure of the mobility of the component and is independent of concentration. At equilibrium, the component will be distributed between the two phases of the system in a given ratio which is independent of concentration at a given temperature. This is known as the Nernst Distribution Law. Strictly speaking, it is the ratio of activities that is truly constant but at very dilute concentrations, as in the food packaging situation, the error will be very small. The distribution constant is often referred to as the partition coefficient.

This law can also be deduced from the concept of chemical potential outlined earlier. At equilibrium, the chemical potential of any component must be the same in both phases. Hence, in phase I:

$$\mu_I = \mu_I^0 + RT\ln a_I$$

and in phase II

$$\mu_{II} = \mu_{II}^0 + RT\ln a_{II}$$

where a_I and a_{II} are the activities of the component in the respective phases. At equilibrium $\mu_I = \mu_{II}$ and as μ_I^0 and μ_{II}^0 are constants at a given temperature

$$\frac{a_I}{a_{II}} = \text{constant}$$

Since diffusion in liquids and solids is a very slow process, in the food packaging situation it is possible that true equilibrium may not have been attained. In practice, much will depend on the time and temperature of storage. However, for periods of six months to one year the distribution of a component between the plastic and the food is likely to be approaching very close to its equilibrium value.

ADSORPTION

The diffusion theory is just one concept in physical chemistry which has been used to further the understanding of the migration phenomenon. An alternative approach considers the situation at the interface or boundary between two different molecular species and investigates the forces acting on the monolayer of molecules trapped near the surface. The properties of this surface layer of molecules are often different from the properties of the molecules in bulk. There are two main types of adsorption process:

(1) physical
(2) chemical—often referred to as chemisorption.

Adsorption is a type of adhesion which occurs at the surface of a solid, or a liquid in contact with another medium, resulting in an increased concentration of molecules in the immediate vicinity of the surface. The adsorbed layer is often only a monolayer in thickness and is retained by van der Waals' forces in the case of physical adsorption or by covalent, polar or other bonding forces in the case of chemisorption. At a given temperature there is a definite relationship between the number of molecules adsorbed onto a surface and the pressure (if a gas), or the concentration (if a solution) represented by the equation:

$$\frac{x}{m} = Kp\frac{1}{n} \qquad \text{(Freundlich)}$$

where x = mass of gas adsorbed by m grams of adsorbing material at pressure p. K and n are constants. This is an empirical equation and it holds only over a limited pressure range. It can be plotted graphically as in Fig. 1 and is usually referred to as the Freundlich isotherm. A plot of $\log x/m$ against $\log p$ gives a straight line.

Langmuir treated adsorption from the point of view of a unimolecular layer of gas molecules on the surface of the adsorbent. If n = number of molecules striking a surface 1 cm^2 per second, and α = the proportion of molecules that adhere to the surface, then αn molecules adhere to each

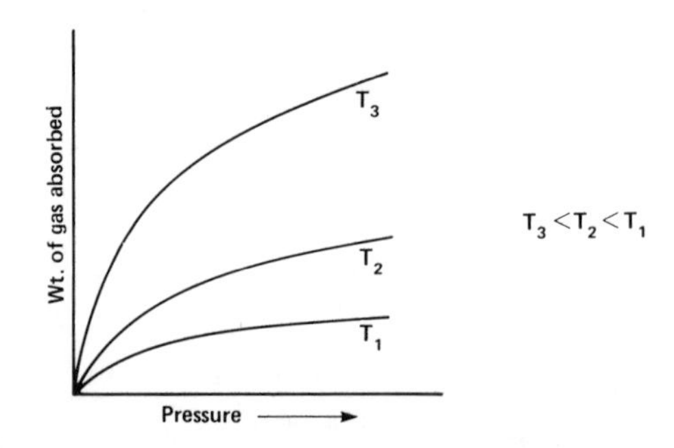

FIG. 1. Freundlich isotherm.

square centimetre of surface per second. If $\theta =$ the fraction of available surface covered with gas molecules at any instant, $1 - \theta$ is the fraction which is bare and $(1 - \theta)\alpha n$ is the actual rate of condensation of gas molecules/cm^2 of total surface.

The evaporation of molecules from the surface is proportional to the area covered and, hence can be represented as $v\theta$ where v is a constant. At equilibrium, rates of evaporation and condensation will be equal.

Therefore,

$$(1 - \theta)\alpha n = v\theta$$

and subsequently,

$$\theta = \frac{\alpha n}{v + \alpha n}$$

If only a monolayer of gas forms on the surface, θ is proportional to the amount of gas adsorbed by the mass m of adsorbent. In addition, n is proportional to the gas pressure, p. Hence,

$$\frac{x}{m} = \frac{K_1 K_2 p}{1 + K_1 p}$$

where, K_1, K_2 are constants.

At low pressures, or for a poor adsorbent, only a small fraction of the surface is covered with molecules. Therefore, θ is very small, and $1 - \theta \simeq 1$. Therefore,

$$\frac{x}{m} \simeq K_1 K_2 p$$

and at high pressures $x/m = K$.

Therefore, the Freundlich isotherm is a special case of the Langmuir equation applicable only over a limited pressure range.

Both isotherms are strictly applicable only to gas/solid interfaces but the theoretical concepts are frequently applied to food packaging situations as well. The significance of Freundlich and Langmuirian isotherms will be discussed later under the Gilbert 'effective zero' concept.

Gibbs Equation

This equation describes quantitatively the adsorption of a *solution* at a surface. For a solution of concentration c,

$$S = -\frac{c}{RT}\frac{\mathrm{d}\gamma}{\mathrm{d}c}$$

where S is the excess *concentration* of solute per unit area of surface, as compared with the concentration in the bulk of the solution, $d\gamma/dc$ is the rate of increase in the surface tension of the solution with the concentration of the solute, R is the gas constant and T is the absolute temperature. Thus, any solute which causes a decrease in the surface tension of the solvent (i.e. $d\gamma/dc$ is negative) will have a higher concentration at the surface than in the bulk of the solution, i.e. it will be adsorbed at the surface. This will be true for many additives in plastic formulations, e.g. slip and wetting agents, which will therefore 'plate-out' onto the surface of the 'solvent', in this case the plastic. Where, however, adsorption from solution leads to the formation of a single layer of solute molecules onto the surface of the adsorbent, the Langmuir equation derived earlier will apply.

MODELS OF MIGRATION PROCESSES

A clear picture of the migration process has been presented by Briston and Katan.[7] The basic model considered interactions between three components: the food, the plastic, and the environment, where the plastic (wrapping or container) is effectively interposed between the food and its environment. In closed containers, the food is essentially isolated from its environment although not all plastics are impervious to gaseous transfer. Hence, tainting may occur and in the reverse direction there may be a loss of volatiles (aroma) from the food. However, migration is concerned only with interaction between the plastic and the foodstuff in which the environment takes no significant part.

Briston and Katan[7] and Knibbe[8] have identified three separate classes of migration.

Class 1—non-migrating.
Class 2—independent migration, i.e. not controlled by the food.
Class 3—migration controlled by food, e.g. leaching, negligible migration in the absence of the food.

Examples of Class 1 include certain dry, hard foods packaged in inert containers, e.g. hard fruit and vegetables, sugar, salt, etc. Other foods at deep frozen temperatures may also approach this ideal. In this situation the diffusion coefficient approaches zero and only the monolayer of component present on the internal surface of the plastic can be dissolved and hence transferred into the foodstuff.

Class 2 systems are illustrated by the diffusion of gaseous components, e.g. VCM from the plastic into the food. In Class 2 migration, the diffusion coefficient possesses a finite value and some transfer of component through the plastic and into the food can occur. However, there is no change in the phase boundary with time. Migration follows Fick's laws of diffusion and the diffusion coefficient is constant and independent of time and the type of food in contact with the plastic.

In Class 3 migration, the food penetrates into the plastic, thus disturbing its physical structure and changing the phase boundary between the food and the plastic. This causes swelling of the plastic, which increases progressively as penetration proceeds. The diffusion coefficient is no longer constant but increases with time, since $D_{\text{swollen plastic}}$ is greater than D_{plastic}. The swollen layer forms an inhomogeneous multi-phase system, for which Fick's laws are inapplicable. There is a gradient of chemical potential across the swollen layer and the mathematical treatment follows that of Rudolph,[5] outlined earlier. Differences between the three classes of migration are illustrated diagrammatically in Fig. 2 although, in practice, such sharp boundaries are seldom so clearly defined.

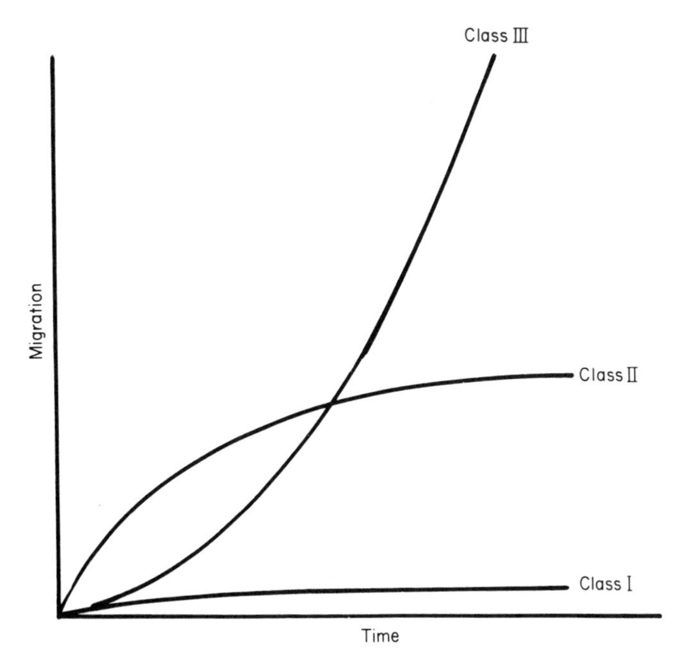

FIG. 2. Diagrammatic representation of different types of migration processes.

'Effective Zero' Migration Concept (Gilbert[9,10])

As an alternative to the diffusion mechanism for migration, Gilbert[9,10] introduced the concept of 'effective zero' migration based on a theoretical treatment of adsorption/desorption processes, as applied to the system PVC–VCM–Food. In his hypothesis, Gilbert argues that the molecular transfer of VCM from the polymer to the food would be retarded by active sites in the polymer, giving rise to adsorption effects on the surface of the polymer at the very low but finite concentrations of monomer now present in the polymer. Thus, the desorption phenomenon would not necessarily be linear. Even though migration of VCM is readily demonstrable at higher monomer concentrations, it does not follow that extrapolation using a linear expression to lower levels is scientifically sound. Monomer levels in PVC are now so low as to be essentially zero. At these concentrations Langmuirian or Freundlich desorption isotherms can be used to define the equilibrium state and distribution of monomer between the polymer and the food.

Some experimental support for this concept has been obtained by measurements on PVC resins containing known amounts of VCM in contact with a measured volume of an extractant (for example, water, hexane, edible oil). At equilibrium, the partition to the resin phase was calculated from the concentration determined in the liquid phase. It was found that for an unplasticised polymer the partition ratio became exponential towards the polymer at low monomer concentrations. The data fitted the Freundlich equation at low monomer concentrations whilst at slightly higher values the fit was more nearly Langmuirian. At higher concentrations still, clustering of VCM molecules is observed around initially bound VCM molecules through intermolecular attraction forces. The amount of clustering is naturally reduced at higher temperatures. In a later investigation (Biran *et al.*[11]) the experiments were repeated using selected food constituents, e.g. corn oil, casein and sucrose. The latter product did not adsorb detectable amounts of VCM under the conditions chosen. Corn oil was found to absorb 10–15 times the amount of VCM adsorbed by water. A study of the mechanism of adsorption of VCM by dry casein suggests that surface adsorption takes place at active sites on the casein particles.

PIRA Migration Studies

The migration of non-volatile additives from plastics has been studied in some depth by PIRA.[12] They used model systems designed so that relatively large migrations took place and the additives were selected for analytical convenience. Magnesium cyclohexylbutyrate (Mg CHB) was incorporated into PVC sheets and polystyrene sheet was loaded with phenolphthalein. Penetration of solvent into the plastics was determined by measurements of weight gain. This work established that there is no significant migration of additive from polymer to solvent without penetration of the polymer by the extractant (Fig. 3). In a second series of experiments, the extractant chosen was a mixture of one polymer-penetrating liquid and one non-penetrating liquid, the additive being soluble in both liquids. In such a system it was shown that migration increases as the proportion of penetrating to non-penetrating liquid increases. However, the non-penetrating component can have a marked effect on the migration depending on its influence on the rate of diffusion of the additive in the plastic. It was also shown that, where the penetrating liquid is also a solvent for the polymer, there is a critical point in the migration curve which correlates with temperature changes that occur when the two liquids are gradually mixed. The critical point marks the start of significant polymer swelling, whereas in the food packaging situation the change in physical properties would mean that it could no longer function

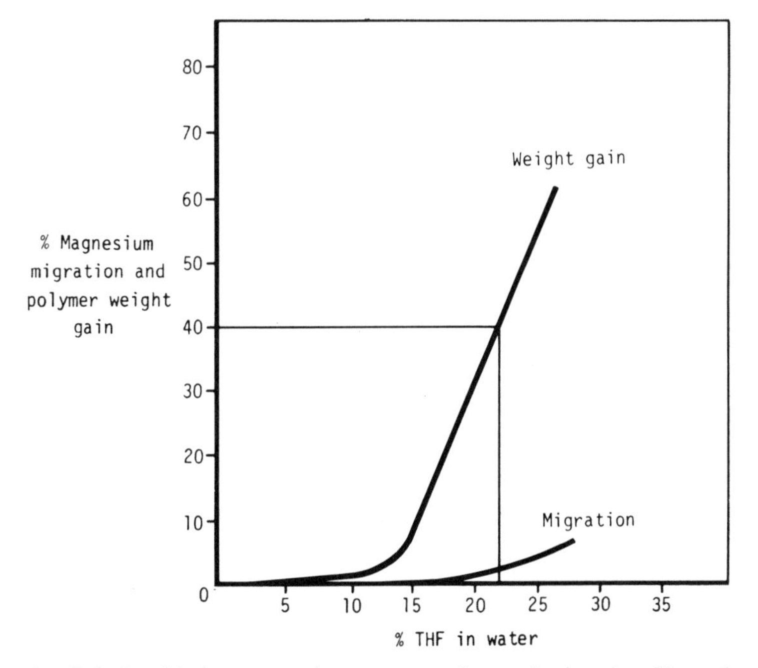

FIG. 3. Relationship between polymer penetration and migration. Reproduced with permission from *Food Manufacture*, May 1979.

as a packaging material. No packaging material that was swollen on contact with a particular food would be acceptable as a contact material. Hence, if the food does contain a constituent that can penetrate into the plastic, the proportion of that constituent must be below the critical point. Therefore, since there is then no swelling, migration will be minimal. As a result of their studies the PIRA team have developed a pictorial concept of migration as an alternative to a mathematical expression of the transport process. Consideration of the migration process in these terms suggests that the fitness of a plastic, for use as a food packaging material, should be judged from the standpoint of specific migration into a suitable mixture of polymer-penetrating and non-penetrating liquids, rather than by empirical measurements of global migration into food simulating solvents.

REFERENCES

1. Garlanda, T. and Masoero, H., Considerations on the migration of plastics components in food-simulating solvents, *La Chimica e L'Industria*, 1966, **48,** 936.
2. Davies, J. T., Migration of styrene monomer from packaging into food. Experimental verification of a theoretical model, *J. Food Technol.*, 1974, **9,** 275.
3. Crank, J., *The mathematics of diffusion*, 2nd edn, Oxford University Press, Oxford, 1975.
4. Robinson, L. and Becker, K., Behaviour of polyolefins in contact with edible oil, *Kunststoffe*, 1965, **55,** 233.
5. Rudolph, F. B., Diffusion in a multi component inhomogeneous system with moving boundaries. I. Swelling at constant volume, *J. Polymer Sci.*, 1979, **17,** 1709.
6. Figge, K. and Rudolph, F. B., Diffusion processes between plastic package and contents, *Angew. Makromol. Chem.*, 1979, **78,** 157.
7. Briston, J. H. and Katan, L. L., *Plastics in contact with foods*, Food Trade Press Ltd, London, 1974.
8. Knibbe, D. E., Theory of extraction of additives from plastics by swelling solvents, *Plastica*, 1971, **8,** 358.
9. Gilbert, S. G., Low molecular weight components of polymers used in packaging, *Environ. Health Perspect.*, 1975, **11,** 47.
10. Gilbert, S. G., Migration of minor constituents from food packaging materials, *J. Food Sci.*, 1976, **41,** 955.
11. Biran, D., Gilbert, S. G. and Giacin, J. R., Sorption of vinylchloride by selected food constituents, *J. Food Sci.*, 1979, **44,** 56.
12. Adcock, L. H., Hope, W. G. and Paine, F. A., The migration of non-volatile compounds from plastics: Part I. Some experiments with model migration systems, *Plastics & Rubber Materials & Applications*, 1980, Feb., 37.

Chapter 7

MIGRATION—EXPERIMENTAL DETERMINATION

INTRODUCTION

Most legislative controls for plastics in contact with foods require evidence as to migration or extractability in order to assess possible contamination of the diet and hazard to the consumer. The theoretical aspects of migration have been discussed previously. Extractability is used here to define the transference of a compound from a plastic to a food, irrespective of whether the mechanism is diffusion, leaching or a combination of the two. This section concentrates on the principles behind the practical measurement of extractability.

In principle, the extractability of a compound from a plastic by a foodstuff can be determined by placing the plastic (of known surface area) in contact with the food under defined conditions of temperature and time. At the end of the test an appropriate analytical technique is used to determine the amount of compound present in the food and, hence, the migration can then be calculated and is usually expressed as mg/dm^2. In practice, the test is not so simple for the following reasons:

(1) the compounds present in the plastic under test may be unknown and may have been degraded during processing,
(2) many such compounds are difficult to determine analytically in a matrix as complex as food, particularly where only small amounts are present in the extract,
(3) compounds other than the one of interest may also be extracted and subsequently interfere with the analytical determination,
(4) most foodstuffs are stable for short periods of time only, whereas extractability data may be required from long-term studies,

(5) appropriate test conditions are not easy to define as a result of the wide variation in possible contact conditions likely to be encountered in practice in warehouses, supermarkets, corner shops and household larders.

If migration and extractability data are to be of any value for legislative control, the test conditions chosen must reproduce the worst possible (i.e. the most extreme) conditions of contact likely to be met in practical use. Although foods may be stored for long periods in contact with plastics, testing over equivalent periods of time is impractical. Hence, accelerated tests at elevated temperatures are employed. Some foods, e.g. milk, cream, unsaturated oils, are not stable at temperatures above ambient for any length of time. Whilst it is always desirable to use foods themselves for extractability testing, in practice it is seldom possible, for the reasons given above, and so food simulants have to be used instead.

FOOD SIMULANTS

Foods consist of a complex mixture of water, fats, proteins and carbohydrates as well as minor constituents such as vitamins, minerals and synthetic compounds added during processing, e.g. colours, antioxidants, preservatives, stabilisers, flavourings, etc. Data on the composition of raw, cooked and processed foods has been compiled by McCance and Widdowson,[1] particularly those aspects which are important to the nutritionist and dietician. However, even this major work does not present the true picture of the complexity of foods as seen by the chemist. For example, the potato is known to contain at least 150 chemical entities despite the fact that 76 per cent of the flesh is water.[2] Nursten and Williams[3] have found 98 volatile components in blackcurrants, including 23 hydrocarbons, 14 carbonyls, 30 alcohols, 8 acids, 22 esters and 1 ether. Forty-two chemical constituents have been identified[4] in orange oil as well as a number of unidentified compounds. Many other examples could be given to illustrate this point. Furthermore, the number of chemicals added to food, in contrast to natural constituents, is equally forbidding. The Chemical Rubber *Handbook of Food Additives*[5] includes a list of 1697 compounds, comprising 30 preservatives, 28 antioxidants, 44 sequestrants, 85 surfactants, 31 stabilisers, 24 bleaching agents, 60 buffering aids, 35 colours, 9 sweeteners, 116 nutrient supplements, 720 flavours, 357 natural colours and 158 miscellaneous additives. Obviously only a fraction of these will be present in any one food. However, it does serve to emphasise the complexity of food composition facing the analytical chemist and the difficulty in selecting a simple food simulant which can accurately reflect the extraction characteristics of such a mixture. One further difficulty is that foods, being natural products, are additionally subject to compositional variations depending on local conditions, such as the weather, nutritional status, time of harvest, etc. Furthermore, most simulants can be given an adequate specification to limit composition variations. Table 1 shows a specification for olive oil, proposed as a fatty food simulant by an EEC Working Party.[6] This would not be easy for processed foods such as salad cream or yoghurt.

TABLE 1

SPECIFICATION FOR OLIVE OIL TO BE USED AS A FATTY FOOD SIMULANT

Iodine number (Wijs)	80–88
Refractive index at 25 °C	1·466 5–1·467 9
Acidity (expressed as per cent oleic acid)	0·5, maximum
Peroxide number (expressed as oxygen milliequivalents per kg oil)	10, maximum

In an attempt to simplify this situation, foods are often classified into groups such as dry solids, neutral liquids, acidic, ethanolic or fatty substances. Extraction data are not usually required for the first category, although transfer has been demonstrated in products such as dehydrated soups, starch or whole milk powders. This may be particularly relevant for dry solids flowing through pipes, for which kieselguhr impregnated with a known weight of test fat HB307 has been suggested as a suitable simulant.[7] Despite these difficulties, a wide measure of agreement has been obtained in the selection of simulants to represent the action of all but fatty foods. Thus, food simulants fall into four separate classes:

A—Distilled water;
B—Dilute solutions of acids;
C—Ethanol/water mixtures;
D—Fatty food simulants.

Distilled water is used to simulate the extraction capabilities of foods with pH values of 5 and above. In some cases it may be more appropriate to use sodium bicarbonate to test plastics in contact with fish, since packed fish can contain high concentrations of ammonia and substituted amines. Acidic foods ($pH < 5$) such as vinegar, pickles or fruit juices are usually represented by dilute solutions of acetic acid. The concentration

TABLE 2
INFLUENCE OF ORANGE OIL ON TIN STABILISER MIGRATION[8]

Extractant	Concentration of organo tin (ppm) after storage at room temperature for: 1 month	2 months	3 months	6 months
Vinegar	0·2	0·4	0·3	0·3
Orange drink (0·1 per cent orange oil)	0·9	1·2	1·0	1·2
Orange drink (0·2 per cent orange oil)	2·3	2·2	3·1	4·3

recommended has varied from 2 to 5 per cent, but it is unlikely to be a critical factor once specified. Citric acid and lactic acid have also been suggested, but the former is unlikely to be an adequate simulant for fruit juices since the essential oils present probably represent the major aggressive constituent. This is illustrated in Table 2, in which orange drink is compared with vinegar as an extractant. Concentrations of ethanol used in aqueous mixtures have varied from 5 per cent for beers and cider, 15 per cent for table wines and up to 50 per cent for spirits (on a v/v basis). Other simulants have included sodium chloride and sucrose. In contrast, for fatty foods the choice is much more difficult and the problem will be considered in more detail below.

CHOICE OF FAT SIMULANT

The difficulty in simulating the migration of a substance from a plastic into a fatty food is reflected in the large number of extractants that have been proposed, e.g. *n*-heptane, diethyl ether, ethanol or liquid paraffin, as well as fats such as coconut oil, olive oil, sunflower oil, lard and various synthetic triglycerides.[9] The first category of simulants, i.e. simple organic solvents, usually exhibit extraction characteristics markedly different from the observed migration of an additive into foods. Even oils and fats themselves are not entirely representative of the behaviour of fatty foods and can present nearly as many analytical difficulties as foods. The extractability of such substances has been most carefully studied by Figge and Koch[9,10] using specially synthesised additives containing labelled carbon atoms.

The additives were incorporated into various plastics and migration was assessed by radiochemical measurements. The compounds selected are shown in Table 3. Migration was assessed both in short-term (5 h at 65 °C)

TABLE 3
RADIOCHEMICALLY LABELLED ADDITIVES USED IN MIGRATION EXPERIMENTS

(1) Di-*n*-octyl[1-^{14}C]-tin-2-ethyl-*n*-hexyl dithioglycollate

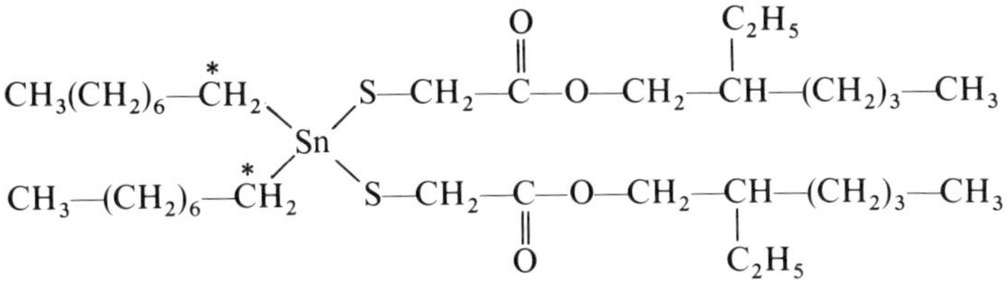

$$\begin{array}{l} CH_3(CH_2)_6-\overset{*}{C}H_2 \\ \qquad\qquad\qquad\qquad \diagdown \\ \qquad\qquad\qquad\qquad\quad Sn \\ \qquad\qquad\qquad\qquad \diagup \\ CH_3-(CH_2)_6-\overset{*}{C}H_2 \end{array} \begin{array}{l} S-CH_2-\overset{O}{\overset{\|}{C}}-O-CH_2-\overset{C_2H_5}{\overset{|}{C}H}-(CH_2)_3-CH_3 \\ \\ \\ \\ S-CH_2-\underset{O}{\underset{\|}{C}}-O-CH_2-\underset{C_2H_5}{\underset{|}{C}H}-(CH_2)_3-CH_3 \end{array}$$

(2) 1,3,5-trimethyl-2,4,6-tris-(3,5-di-*tert*-butyl-4-hydroxybenzyl[^{14}C]benzene)

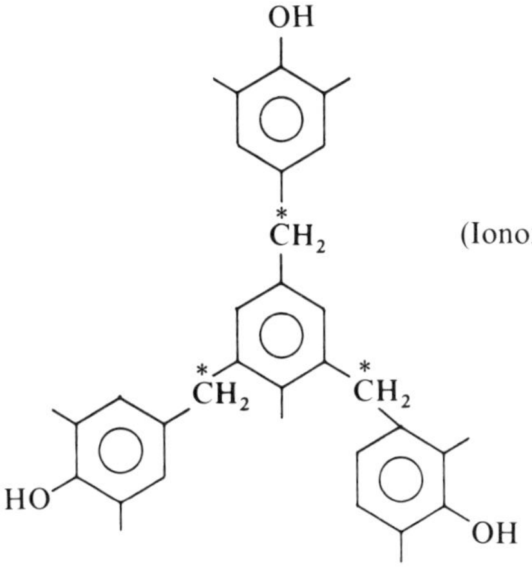

(Ionox 330)

OH, $\overset{*}{C}H_2$, $\overset{*}{C}H_2$, $\overset{*}{C}H_2$, HO, OH

(3) *n*-butyl stearate[1-^{14}C]

$$CH_3-(CH_2)_{16}-\underset{O}{\underset{\|}{\overset{*}{C}}}-O-CH_2-(CH_2)_2-CH_3$$

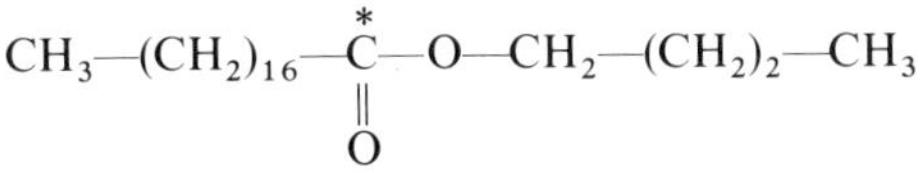

(4) Stearic acid[1-^{14}C]amide

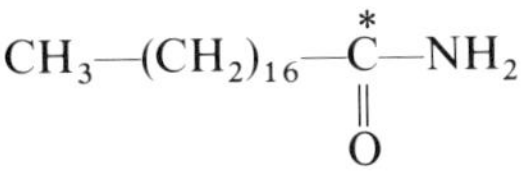

$$CH_3-(CH_2)_{16}-\underset{O}{\underset{\|}{\overset{*}{C}}}-NH_2$$

and long-term (30 or 60 days at 20 °C) tests. The results obtained in the study of the migration of Ionox 330 from HDPE are shown in Fig. 1. Similar discrepancies were observed with the organotin stabiliser in PVC, in particular, and with the other additives and plastics too under certain conditions. Diethyl ether gave extremely high extraction values compared to sunflower oil in all the additive/plastics combinations tested. Paraffin oil did not extract measurable amounts of antioxidant from HDPE, whereas *n*-heptane dissolved more than 50 per cent within 5 h at 65 °C. However,

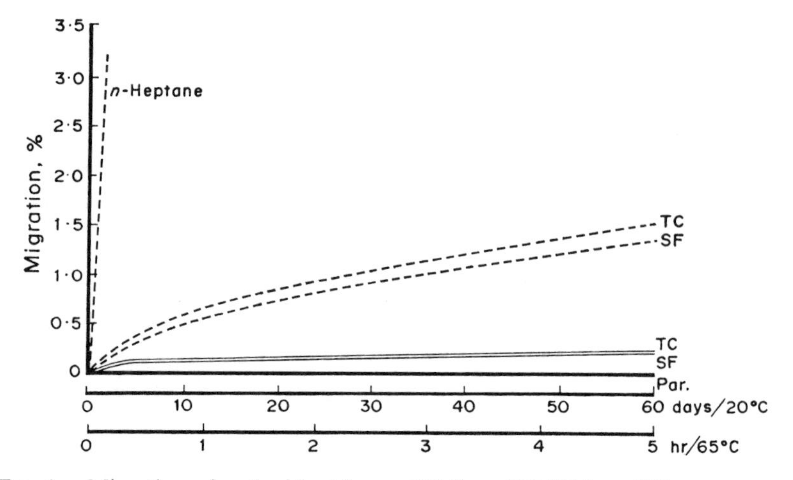

FIG. 1. Migration of antioxidant Ionox 330 from HDPE into different contact liquids as a function of time. Tests were carried out at 20 °C for 60 days (——) and at 65 °C for 5 h (----) using sunflowerseed oil (SF), tricaprylin (TC), paraffin oil (Par) and *n*-heptane.

whilst *n*-heptane appeared to be a suitable simulant with PVC, tricaprylin caused excessive swelling of the plastic, with the result that the degree of extraction was 250 times higher than that observed with sunflower oil. A range of higher alkanes and di-*n*-alkyl ethers was evaluated but all were found to be unsuitable, since it was impossible to derive correlation factors for the amounts migrating into edible oils from the amounts observed to migrate into such solvents. Thus, simple organic solvents tend to extract far too great a proportion of the additive and cause excessive swelling (almost solubilisation in some cases) of the plastic or lubricant, particularly at high temperatures. The high extractability of organic solvents for plastic components compared with that of edible fats would not be so important if the components were extracted in the same quantitative ratio by the test liquids and by edible fats, since it would then be possible to calculate the true extent of migration using correction factors. No organic solvents meet this objective. For example, compounds (2) and (4) (Table 3) migrate into both sunflower oil and into tricaprylin in a ratio of 1:8, compared to 1:1 for *n*-heptane and 1:6 for di-*n*-butylether. Since there is such a difference in chemical structure between these organic solvents and natural edible oils and fats, it is hardly surprising that they exhibit markedly different interaction behaviour towards packaging materials and so can never be used as a true fat simulant.

Edible oils are less objectionable on these grounds. However, they do present a number of analytical difficulties and considerable variation in extraction behaviour has also been observed. The unsuitability of tricaprylin in contact with PVC has already been referred to. This led Figge[9] and his team to develop an analytically pure synthetic triglyceride for use as a fat simulant. They found that the amount of additive extracted by a triglyceride from HDPE, PVC and PS was a function of the chain length of the acyl residue, passing through a maximum as shown in Fig. 2. For HDPE, a broad maximum in the range of five to ten acyl carbon atoms was observed with a slight fall-off for longer chain fatty acids. For PVC and PS, the maximum was much sharper and occurred at around three to five carbon atoms with a sharp fall-off for longer chain compounds. Hence, Figge[9] prepared a synthetic triglyceride (Fettsimulans HB307) which was a mixture of synthetic triglycerides with a fatty acid composition similar to the strongly extracting coconut oil. The composition is shown in Table 4. It contains 50 per cent lauric acid (C_{12}) with 20·5 per cent of lower and 29·5 per cent of higher fatty acids. The synthetic product possesses a number of characteristics that are useful analytically, e.g. low unsaponification value, low mono and diglyceride content, high saturation for good stability (low

TABLE 4

COMPOSITION OF FAT SIMULANT HB307

Fatty acid distribution										
Number of carbon atoms	6	8	10	12	14	16	18			
GLC area, per cent	0·5	7·5	10·3	50·4	13·9	7·8	8·6			
Glyceride distribution										
Number of carbon atoms	22	24	26	28	30	32	34	36	38	40
GLC area, per cent	0·1	0·3	1·0	2·3	4·9	10·9	13·9	21·1	16·1	11·7
Number of carbon atoms	42	44	46	48	50					
GLC area, per cent	9·8	4·4	2·2	1·1	0·2					

The iodine value (Wijs) $\leq$ 0·5, acid value = 0·02 and the m.pt. = 28·5 °C.

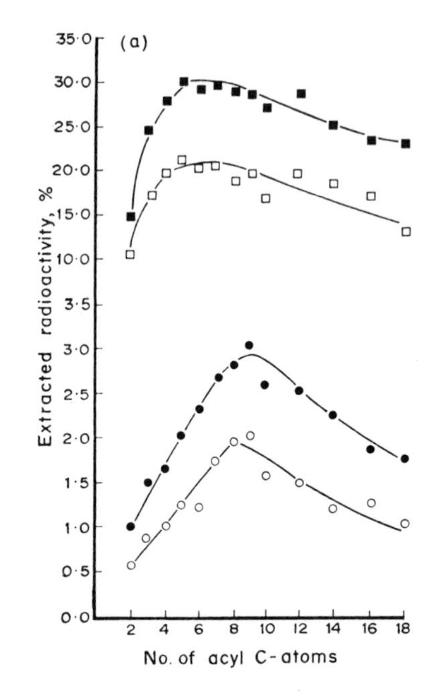

FIG. 2. Dependence of the amount of radioactivity or additive extracted on the chain length of the acyl residues in triglycerides. Levels of extraction were determined for (a) stearic acid amide (□, ■) and Ionox 330 (○, ●) from HD-PE, (b) Advastab 17 MOK (△, ▲) and Ionox 330 (▽, ▼) from PVC and (c) *n*-butyl stearate (◇, ◆) and Ionox 330 (□, ■) from PS after 2 h (open symbols) and 5 h (solid symbols) at 65 °C.

iodine value), and good transmission in the UV region of the spectrum. The melting point is 28 ± 2°C. It does, however, give a GLC chromatogram with many peaks, whereas sunflower oil produces only four major peaks and is easier to use.

Comparative migration tests using HB307, coconut oil, groundnut oil (partially hydrogenated) and butter were then carried out. Butter contains a large number of short chain fatty acids whilst groundnut oil contains many long chain compounds. The results of the migration tests are shown in Table 5. It can be seen that there is a reasonable agreement between the observed migration into the synthetic triglyceride mixture HB307 and that observed into the other fatty materials. HB307 gave slightly higher values than the other fats and so fulfils the requirement that the most extreme

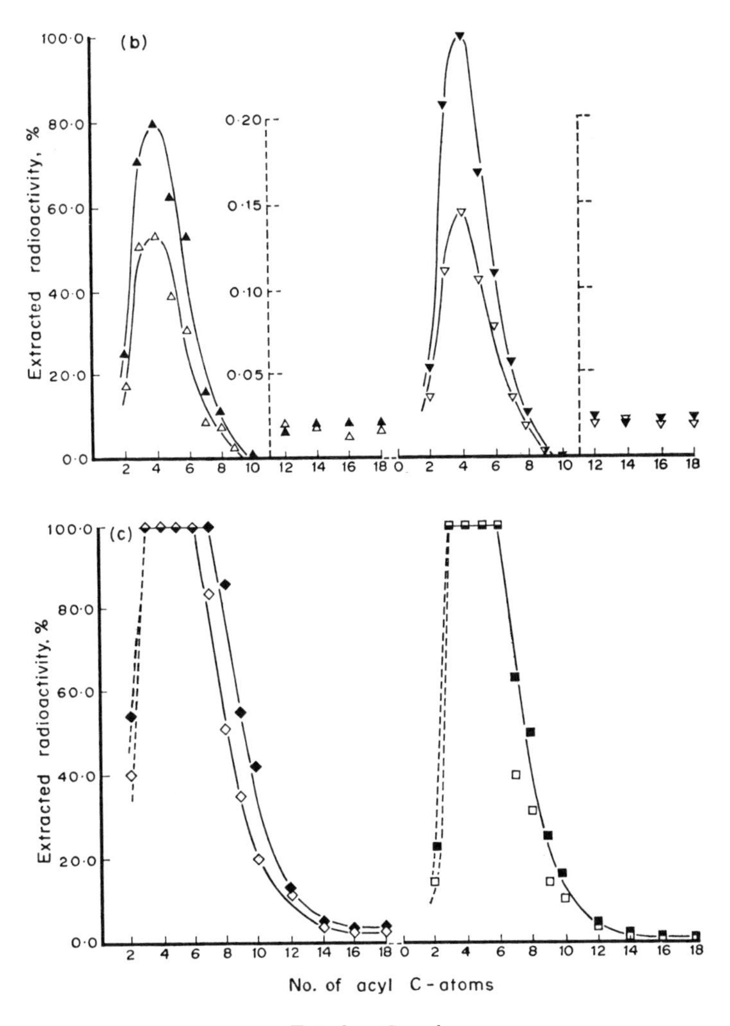

FIG. 2. *Contd.*

conditions are assessed. More recent migration studies[11] have compared test fat HB307 with beef tallow, lard, fish oil and margarine. Only small deviations in migration behaviour between HB307 and the other fats have been observed, suggesting that this synthetic product is the best fat simulant available when foodstuffs themselves cannot be used.

TABLE 5

COMPARISON OF THE MIGRATION OF ADDITIVES FROM PLASTICS INTO EDIBLE OILS AND A STANDARD TRIGLYCERIDE MIXTURE AFTER 60 DAYS AT 20 °C

Plastic (with labelled additive, see Table 3)	Per cent migration			
	Groundnut oil	Coconut oil	Butter	Triglyceride mixture
PVC + compound (1)	0·009	0·014	0·017	0·016
HDPE + compound (2)	0·090	0·098	0·120	0·140
HDPE + compound (4)	0·80	0·96	1·05	1·36
PS + compound (2)	2·08	2·53	3·07	3·05
PS + compound (3)	5·20	5·61	7·11	7·57

INFLUENCE OF TIME AND TEMPERATURE ON MIGRATION

Increasing dependence on convenience foods and storage in coldrooms, freezers, and domestic refrigerators as well as boil-in-bag type applications has led to a need for migration data over a wide temperature range. The need for data to be available without excessive delay requires the use of accelerated tests at elevated temperatures. Van der Heide suggested that a test for 10 days at 45 °C was roughly equivalent to six months storage at 25 °C. This proposal has been largely confirmed by Woggon *et al.*[12] However, most workers have followed the suggestion of Hofmann and Ostromow[13] and used 10 days at 40 °C. Woggon *et al.*[14] proposed a series of tests to cover the simulation of various storage/heat treatment conditions as shown in Table 6. Currently recommended testing times and temperatures are shown in Chapter 5, Table 5.

Systematic studies by Figge and Koch[10] of the migration of *n*-butyl

TABLE 6

TEST CONDITIONS TO SIMULATE MIGRATION UNDER REAL TIMES AND TEMPERATURES

Real time/temperature	Simulating conditions
Storage for nine months at −18 °C	48 h at 45 °C
Storage for six months at 25 °C	10 days at 45 °C
Bottling at 70 °C	1 h at 70 °C
Pasteurisation in the packet	$\frac{1}{2}$ h at 100 °C
Sterilisation in the packet	$\frac{1}{2}$ h at 120 °C

stearate from PS into HB307, in the temperature range 30 to 80 °C, showed that at 40 °C there was no further migration after about three days (Fig. 3). Hence, a 10-day test is adequate to measure the final value of migration. The effect of different temperatures on migration over a fixed period of 10 days is illustrated in Fig. 4. At 40 °C and below, equilibrium conditions are established within a few days. At temperatures of 50 °C and above, however, a sudden increase in migration occurs and equilibrium conditions are nowhere near established within a 10-day period. Hence, unrealistic values may be obtained in such tests, where excessive swelling occurs. Tests at such high temperatures are usually carried out for periods of hours rather than days, and thus simulate more exactly real conditions of use.

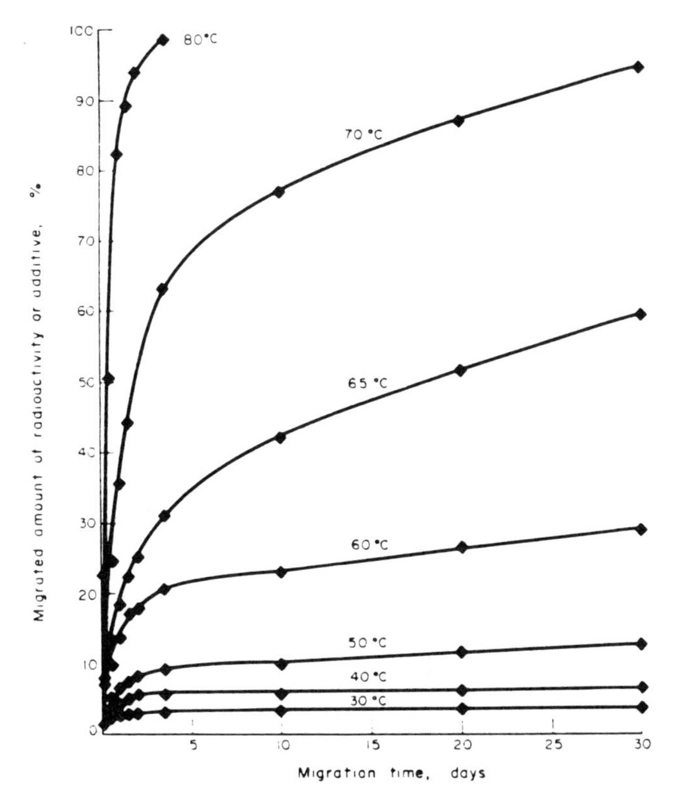

FIG. 3. Migration of *n*-butyl stearate [1-^{14}C] from polystyrene film into fat simulant HB 307 during periods of 1–30 days at temperatures between 30 and 80 °C.

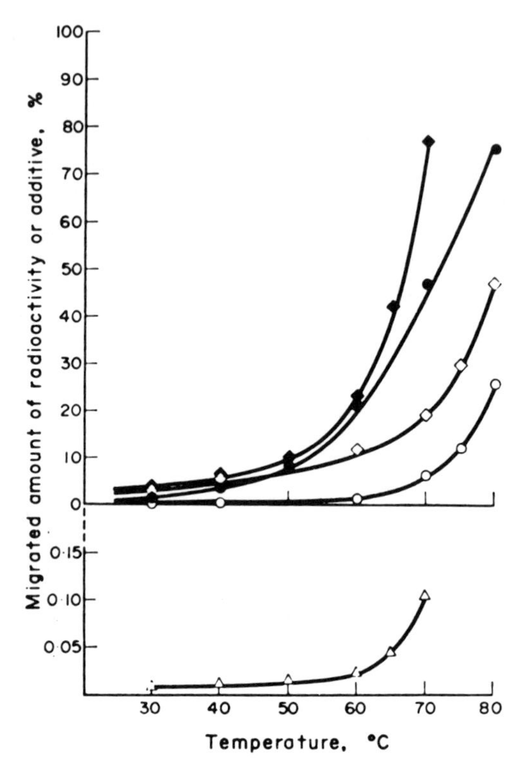

FIG. 4. Migration of labelled additives (Ionox 330 (○—○) and stearic acid amide (●—●) from polyethylene film, Ionox 330 (◇—◇) and *n*-butyl stearate (◆—◆) from polystyrene film Irgastab 17 MOK (△—△) from PVC film) into fat simulant HB 307 during a 10-day period at temperatures between 30 and 80 °C.

GLOBAL MIGRATION (GM)

The concept of a global migration (or total migration) test was introduced:

(1) as a control on the transference of undesirable, non-nutritive chemicals into food, whether they have any toxicological significance or not, and;
(2) to avoid having to test separately for each specific migrating compound that may be present (in many cases the compounds present are unknown).

In principle, the measurement of total migration from a packaging material is very simple. A sample of the material under test, of known surface area, is placed in contact with a food simulant under specified conditions of temperature and time. At the end of the test, the simulant is evaporated and the residue is dried and weighed. Aqueous simulants are often extracted with an immiscible organic solvent such as chloroform which, in turn, is evaporated and the residue dried and weighed. The result is then expressed as mg of residue/dm^2 surface area of material in contact with the simulant.

This approach is generally satisfactory for simulants A, B and C, apart from the fact that any steam volatile materials, azeotropes and volatile organic compounds, e.g. VCM will be lost during the evaporation stage. Although few interlaboratory tests have been carried out, the method should be reproducible to about $\pm 1\ mg/dm^2$. This method cannot, of course, be used for fat simulants except where simple organic liquids are used.

Fat Test

In the case of fat simulants such as olive oil, coconut oil, HB307 or sunflower oil, etc., the evaporation technique cannot be used. In this test, samples of plastic of known area (S) are weighed before (W_1) and after (W_2) contact with the fatty simulant. The global migration is then given by:

$$GM = \frac{W_1 - W_2}{S}$$

In practice, W_2 is usually greater than W_1, despite migration of compounds from the plastic into the oil, since a greater weight of oil is absorbed by the plastic than the weight of compound leached out of the plastic. Hence, the change in weight as calculated above has to be corrected and an allowance made for the weight of oil (F) absorbed by the plastic. Then:

$$GM = \frac{W_1 - W_2 + F}{S}$$

A number of different procedures have been proposed for the determination of global migration. They differ chiefly in the simulant chosen and in the method used to measure F. The basic distinguishing features of each method will now be described.

Methods for the Determination of GM

Community method[6]—A method developed by an EEC Working Group which uses olive oil as the fatty simulant. Absorbed oil is determined by GLC, measuring the height of the methyl oleate peak and using margaric acid or the triglyceride as the internal standard.

Pallière method[15]—Sunflower oil is used as the simulant and extracted oil is determined iodometrically. This method is simpler and quicker than most other methods based on GLC or radiochemical measurements but it is subject to interference from unsaturated co-extractants leached from the plastic by the oil. Although a 'blank' test is specified, the correction is not always easy to apply. The iodometric technique cannot be applied to saturated oils or triglycerides.

Van Battam and Rijk method[16]—This uses HB307 as the simulant and the fat is extracted and determined by GLC using hydrocinnamic acid as the internal standard and measuring the methyl laurate peak—the principal fatty acid (C_{12}) occurring in HB307 (up to 50 per cent).

Koch and Kröhn method[17]—This again uses HB307 but GLC is used to measure the triglyceride C_{36} peak with undecanoic acid triglyceride as the internal standard. Alternatively, the fat can be estimated by an enzymatic measurement of the glycerol present.

Figge method[18]—This uses a radio-actively labelled form of HB307 in which ^{14}C is evenly distributed amongst all the fatty acids and glycerol. Absorbed triglyceride is extracted and then measured by liquid scintillation counting.

Rossi et al.[19] *method*—Sunflower oil was used as a simulant and the residual oil was determined by GLC, measuring the linoleic acid peak ($C_{18:2}$).

Piacentini[20] experimented with coconut and olive oils, measuring the extracted fat by GLC after methylation or silylation.

PRACTICAL PROBLEMS IN THE DETERMINATION OF GLOBAL MIGRATION

Sample Holders

Plastics under test can vary from very thin flexible films to fairly thick and rigid specimens. The former are difficult to keep apart from each other when immersed in the test liquid, where they must not be allowed to touch. Rigid specimens may bend, buckle or tear during immersion, and small fragments may be lost producing errors in the true weight change. A number of different holders have, therefore, been developed (see Fig. 5). The 'toast

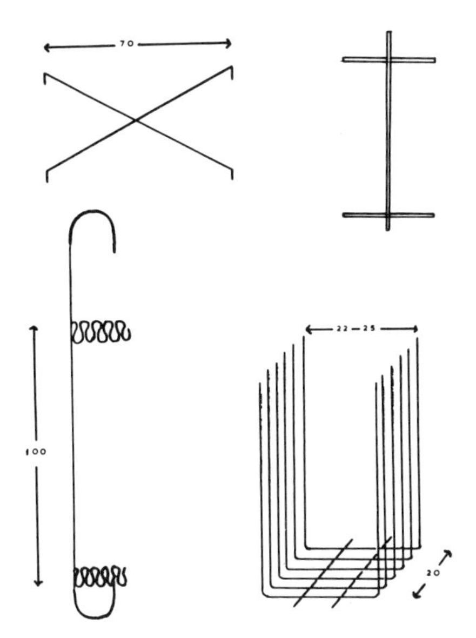

FIG. 5. Sample holders.

rack' is preferred by many workers for rigid samples, whilst films are usually wrapped in a wire mesh roll. The holder shown top left in Fig. 5 is often used where the experiments are conducted in petri dishes. For laminates, where only one side is to come into contact with the simulant, special Maturi cells have been designed.[21] Alternatively, the plastic can be made up in the form of a bag.

Atmospheric Conditioning

Some samples gain or lose moisture with changes in relative humidity of the atmosphere. Since the determination of GM relies on the measurement of very small changes in weight during immersion in a test fat, it is important that this source of error is eliminated. Hence, all plastics should first be weighed and then conditioned in an atmosphere of 50 per cent relative humidity, at 20 °C. A desiccator containing 35 per cent (v/v) sulphuric acid provides a suitable atmosphere. The conditioning is repeated after immersion in the test fat. Conditioning is particularly important for plastics such as ABS, NBR and polyamide. In some methods it is recommended that the sample should first be conditioned at 80 per cent relative humidity (20 per cent v/v sulphuric acid). After 24 h the sample is weighed and transferred to a second desiccator at 50 per cent RH. After a further 24-h period the sample is reweighed. If the change in weight is less than 1 mg, no conditioning is required. Otherwise, the sample should be conditioned at 50 per cent RH until two successive weighings taken at 24-h intervals differ by less than 1 mg. The conditioning is repeated after contact with the test oil and the stable weight at 50 per cent RH is used in the calculation of migration.

Determination of Oil Absorbed

This is the most crucial step in the whole method. Errors can arise from two sources:

(1) incomplete extraction of residual fat from the plastic;
(2) error in the GLC determination of oil in the extract.

(1) *Solvent extraction*—Following immersion in the test fat, excess oil is allowed to drain from the plastic and holder. In some cases the plastics are removed from the holders, blotted to remove excess oil using a photographic type roller (which presses the plastic inbetween sheets of filter paper), reweighed (after conditioning) and then cut into small pieces prior to extraction of the oil with a suitable solvent. At other times, samples and holders are placed directly in the extraction apparatus. Most workers have preferred to use trichlorotrifluoroethane (freon, b.pt. = 47·6 °C) as extraction solvent in a Soxhlet apparatus. The extraction must be continued for at least 6 h, ensuring that the plastic does not float in the apparatus. Most plastics will require a second 6-h extraction. It may be more convenient to carry out the extraction overnight for about 18 h as freon is a non-inflammable solvent. In any case, the analyst must ensure that the extraction of residual oil has been carried out to completion. With samples such as ABS and PS the above process is insufficient and alternative measures must be taken (see later).

(2) *Gas chromatographic determination of residual oil*—After extraction of the residual oil by freon, the excess solvent is removed by evaporation. The oil is then saponified and the free fatty acids are methylated with boron trifluoride and methanol and the mixture is then subjected to GLC. Each of these processes if not carried out to completion can introduce errors into the calculation of residual fat. Variations in the volume of extract injected into the gas chromatograph and detector response are reduced by the use of an internal standard, added to the extract before saponification. Where a triglyceride (e.g. glyceryl trimargarate) is used as an internal standard,

it acts as a check on the saponification, methylation and injection stages of the method. Margaric acid and hydrocinnamic acid have also been used as internal standards but they serve only as a check on the methylation and injection stages of the method. The residual oil is then calculated from the ratio of the heights of the methyl oleate/methyl margarate peaks in the case of olive oil, and the ratio of methyl laurate/methyl hydrocinnamate peaks for HB307. In practice this stage of the method is reproducible to better than 5 per cent as shown in Table 7.

In this experiment, glyceryl margarate and margaric acid were added to six flasks each containing a known weight of olive oil. The 12 solutions were

TABLE 7

REPRODUCIBILITY AND RECOVERY OF OLIVE OIL BY GLC WITH INTERNAL STANDARDS

Solution	Weight of oil taken (mg)	Weight of oil found (mg)	Recovery (per cent)	Mean	Standard deviation
(A) Peak Area Measurements:					
Glyceryl trimargarate 1	41·0	41·4	101·0		
Glyceryl trimargarate 2	41·7	41·5	99·5		
Glyceryl trimargarate 3	43·6	42·4	97·2	101·05	3·042
Glyceryl trimargarate 4	60·1	59·7	99·3		
Glyceryl trimargarate 5	64·2	67·3	104·8		
Glyceryl trimargarate 6	65·9	68·9	104·5		
Margaric acid 1	42·7	41·3	96·7		
Margaric acid 2	62·1	62·1	100·0		
Margaric acid 3	65·4	62·7	95·9	96·80	2·273
Margaric acid 4	55·4	52·0	93·9		
Margaric acid 5	62·3	61·6	98·9		
Margaric acid 6	49·9	47·6	95·4		
(B) Using Calibration Graph:					
Glyceryl trimargarate 1	41·0	39·7	96·8		
Glyceryl trimargarate 2	41·7	39·7	95·2		
Glyceryl trimargarate 3	43·6	42·0	96·3	97·867	2·673
Glyceryl trimargarate 4	60·1	58·6	97·5		
Glyceryl trimargarate 5	64·2	66·0	102·8		
Glyceryl trimargarate 6	65·9	65·0	98·6		
Margaric acid 1	42·7	41·5	97·2		
Margaric acid 2	62·1	61·6	99·2		
Margaric acid 3	65·4	64·4	98·5	97·433	1·412
Margaric acid 4	55·4	52·7	95·1		
Margaric acid 5	62·3	60·8	97·6		
Margaric acid 6	49·9	48·4	97·0		

then saponified, methylated and analysed by GLC. The recovery of olive oil was calculated in two ways:

(1) from the total peak areas of the fatty acid esters;
(2) from the ratio $C_{18:1}/C_{17:0}$ peak heights by calibration graph.

The results shown in the table indicate that there is no significant difference between the reproducibility, or recovery, achieved with either internal standard, although as stated above the triglyceride is to be preferred since it acts as a check on both the saponification and the esterification stages of the method. For the determination of global migration from plastics, the use of a calibration graph is essential, since method 1 using the total peak areas of the fatty acid esters is subject to error where components from the plastic are present in the chromatogram.

Where the residual oil extracted exceeds 200 mg, it may be necessary to adjust the volumes of reagents used in subsequent stages of the method and to add further quantities of internal standard. Since only a small portion of the final solution is injected onto the GLC column, any error at this stage is multiplied by the following factor:

$$\frac{\text{Total volume of solution}}{\text{Volume injected for analysis}}$$

Typically, this factor can be of the order of 10 000.

(3) *Incomplete extraction of residual oil*—Experience with global migration methods, in which the residual oil is extracted with a solvent, has shown that certain plastics, notably ABS and PS, produce negative values for the total migration. This can only mean that all the residual oil has not been extracted from the plastic despite the fact that further treatment with solvent produces no increase in the residual oil figure. This conclusion is supported by results on the same samples obtained using the Figge method,[18] in which the residual fat is determined by dissolution of the plastic and scintillation counting. By this approach, positive figures for GM are obtained. As a solution to this problem, Van Battum[22] suggested that the plastic should be dissolved in chloroform and then precipitated successively with methanol. In this way it was hoped that residual oil occluded in the plastic would be more readily extracted.

This approach was evaluated in the author's laboratory by dissolving a 2 dm^2 sample of polystyrene in 100 ml of boiling chloroform. Fifty milligrams of olive oil and 20 mg glyceryl trimargarate (as internal standard) were added. After mixing, the plastic was precipitated with

TABLE 8
RECOVERY OF OCCLUDED OLIVE OIL FROM POLYSTYRENE BY SUCCESSIVE PRECIPITATION STEPS

Step	Recovery of oil (per cent)	Cumulative recovery (per cent)
1	34	34
2	26	60
3	11	71
4	9	80

With Heptane Wash

Standing time	Recovery of oil (per cent)	With blank correction (per cent)
Overnight	114	109
Overnight	113	108
90 min	113	103

100 ml of methanol. After standing, the residue was filtered and the filtrate concentrated to about 50 ml. Residual plastic was again removed by filtration. Residual olive oil was then estimated as in the 'Community method' by saponification, methylation and GLC. The dissolution and precipitation step was then repeated three more times. The recovery of olive oil at each step is shown in Table 8.

Hence, it would appear to be necessary to carry out at least six dissolution/precipitation operations to ensure complete recovery of absorbed olive oil from polystyrene. Further difficulties arise from the low solubility of glyceryl trimargarate in hot methanol; precipitation occurs on cooling. Hence, a modified dissolution procedure has been proposed in which the precipitated polystyrene is washed with heptane during the filtration stage. This greatly improves the recovery of olive oil from precipitated polystyrene. A single precipitation step gave a recovery just over 100 per cent compared with only 34 per cent (Table 8) without the heptane wash. Recoveries greater than the theoretical quantity indicate that some loss of internal standard has occurred. A separate experiment on the same plastic without the addition of olive oil suggests that a 'blank' correction should be made, as shown in Table 8.

Calculation of GM

Having discussed a number of practical difficulties and sources of error in the experimental measurement of GM, I now want to run through a typical calculation to illustrate how small changes in the residual fat determination, in particular, can affect the final result.

The equation for GM was given earlier as:

$$GM = \frac{W_1 - W_2 + F}{S}$$

Pieces of plastic are cut of such a size as to give a 2 dm^2 contact area. Hence, $S = 2$. Experimentally obtained results from a typical experiment were as follows:

$$W_1 = 20{\cdot}2631\ \text{mg};\ W_2 = 20{\cdot}2962\ \text{mg} \quad \text{and} \quad F = 49{\cdot}0\ \text{mg}$$

Therefore:

$$W_1 - W_2 = -33{\cdot}1\ \text{mg}$$

and substituting this value into the equation for GM:

$$GM = (49 - 33{\cdot}1)\tfrac{1}{2} = 8\ \text{mg/dm}^2$$

Suppose that errors in weighing, from say moisture changes, had occurred and that the true weight difference $W_1 - W_2$ should have been 31 mg, i.e., an error in each weight of 1 mg. This would result in a change in GM of only 1 mg/dm^2. This, therefore, is not a major source of error since in practice weighings should be made with less error than assumed in the above calculation. Now consider the possibility of error in the determination of residual oil by GLC. The calculation is based on the ratio of peak heights of the methylesters of oleic acid and margaric acid. For quantitative work it is more usual to measure the peak areas and this involves a measurement of peak width at half height. In a typical experiment the following measurements were recorded:

Oleic acid—peak height 90 mm, width at half height 2 mm = 180 mm^2
Margaric acid—peak height 75 mm, width at half height 3 mm = 225 mm^2

Hence, the ratio $C_{18:1}/C_{17:0} = (180/225) = 0{\cdot}8$

This figure is then used to calculate the weight of olive oil from a calibration graph. Suppose, however, that the true measurements should have been as follows:

Oleic acid—peak height 90 mm, width at half height 3 mm = 270 mm^2

giving a $C_{18:1}/C_{17:10}$ ratio of $(270/225) = 1{\cdot}2$

This is a 50 per cent increase in the weight of residual oil and in the calculation given above, would increase GM to 20 mg/dm^2. A 1·0 mm change is not much more than the thickness of the pen-lines on some recorders. Whilst modern integrators and data processors are not subject to such errors, variations in gas chromatographic parameters (temperature and gas flow fluctuations) could easily result in changes in peak size or shape to a similar extent. Errors in the method arise from the need to measure small differences in relatively large numbers, which are themselves subject to variations of the same order as the difference to be measured.

Change in Composition of the Simulant During the Test

The validity of migration tests using sunflower oil or olive oil has been contested.[23] This follows the observation that the composition of the residual oil extracted from the plastic, at the end of the test, was different from that of the pure oil at the start of the test. Only minor changes in glyceride composition have been observed with HB307, whereas significant changes in the fatty composition of sunflower oil after contact for 10 days at 40 °C are illustrated in Table 9. Similar experiments using olive oil are reported in Table 10. Significant changes were observed after contact with polyethylene, ABS and polystyrene. However, the $C_{18:1}$ oleate peak is only slightly modified so that use of this peak to calculate residual oil would produce a small error only in the value of GM. In the case of ABS and PS, there is a noticeable contribution from the plastic itself. Table 10 also shows the percentage composition of fatty acids in the extracted oil after making a correction for non-oil components extracted from the plastic. Thus for ABS, little change in triglyceride composition occurs once this correction

TABLE 9

CHANGES IN THE COMPOSITION OF SUNFLOWER OIL AFTER PENETRATION INTO VARIOUS PLASTICS

	Per cent of fatty acid			
	C_{16}	$C_{18:0}$	$C_{18:1}$	$C_{18:2}$
Sunflower oil (before test)	7·0	4·5	23·6	64·9
Oil extracted from:				
Polythene	7·8	5·4	25·3	61·5
Polypropylene	9·1	9·6	22·3	58·9
ABS	23·0	18·2	16·9	41·9
Polystyrene	11·4	17·6	19·8	51·3
Polyamide	12·1	17·0	20·3	51·5

TABLE 10
CHANGES IN THE COMPOSITION OF OLIVE OIL AFTER PENETRATION INTO VARIOUS PLASTICS

	Per cent of fatty acid				
	$C_{16:0}$	$C_{16:1}$	$C_{18:0}$	$C_{18:1}$	$C_{18:2}$
Olive oil (before test)	12·7	0·6	2·2	71·9	12·6
Oil extracted from:					
Polythene (MI = 7)	17·7	0·7	3·2	74·1	4·3
ABS	13·9	0·7	3·2	69·1	13·1
Polystyrene	26·9	7·1	27·9	28·9	9·1
ABS[a]	13·1	0·6	1·3	71·5	13·5
Polystyrene[a]	20·7	6·1	3·7	67·1	2·4

[a] After correction for non-oil components extracted. Note: In all cases the figures given are the means of at least six determinations.

factor has been applied. In general, where there is a significant change in triglyceride content, there is a relative increasing enrichment with decreasing chain length of the saturated fatty acids. Unsaturated fatty acid glycerides penetrate less readily into the plastic.

Changes in composition of HB307 after contact with various plastics have also been observed. However, since HB307 is comprised of over 50 per cent of lauric acid (C_{12}), which changes very little on contact with plastic, the error involved in using this peak for the calculation of residual oil content will be very small. Hence, in any test for global migration the composition of the fatty simulant must be determined at the start and at the end of the test. Where the residual fat is estimated by measurement of a single constituent of the fatty acid mixture, it is vital to ensure that there is no enrichment of this constituent through selective absorption by the plastic. Where this cannot be avoided, the radiochemical method of Figge[18] must be used for the determination of residual fat.

COLLABORATIVE STUDIES OF GLOBAL MIGRATION METHODS

The literature on global migration abounds with data that is largely irrelevant since:

(1) there is often little connection between the data and the original composition of the packaging material;

(2) test conditions, e.g. simulant, time and temperature of exposure vary and are not directly comparable;
(3) evidence of reproducibility is often lacking;
(4) and the composition of the plastics used is not fully stated, is not comparable and will vary from batch to batch.

It is of interest, therefore, to summarise the work of an EEC Working Party established to conduct interlaboratory evaluations of several methods for the determination of global migration. The results have been published by Rossi.[6] The plastic materials studied were LDPE, PP, PS, PAM, PVC and ABS. In the first collaborative test, the methods of Pallière,[15] Van Battum[16] and Rossi *et al.*[19] were compared. The Pallière method was rejected since it was found to be subject to interference from unsaturated substances co-extracted from the plastic, although it was quicker and simpler than methods based on GLC. As a result of this work, a new standardised procedure known as the Community method[6] was developed for use in the second trial. This new method was found to be generally satisfactory for a number of polymers but not for ABS. In the third series of interlaboratory tests, the Figge method[18] was included and further work was carried out on the optimum conditions for extraction, as well as a check on the fatty acid composition of the oil before and after contact with the plastic. The Figge method was found to be the most reliable of all the methods tested but it does require specialised, expensive equipment and experienced personnel to carry out the test. The Van Battum method was less reproducible than the Community method and gives slightly higher migration values for the same sample of plastic. Confidence limits were estimated from the variations observed between laboratories. For a migration limit of 10 mg/dm^2, a tolerance of 4·9, 3·5 or 2·8 mg/dm^2 would be required with the Community method, depending on whether the plastic was examined in a single test or by an average of two or three measurements.

SPECIFIC MIGRATION

This is the study of the movement of single chemical compounds from the plastic into a foodstuff or food simulant. The chemicals of interest as additives are usually complex molecules (Table 3) and difficult to determine analytically at low concentrations, especially in foodstuffs or fatty food

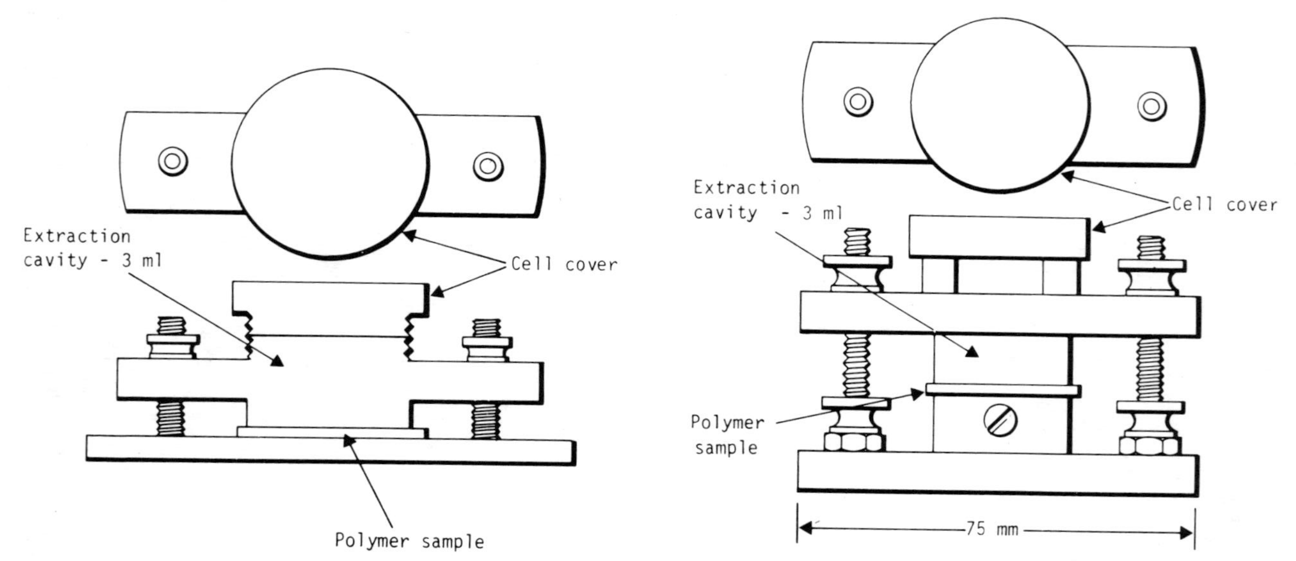

FIG. 6. (Left) A single cavity cell, for measuring migration from one face of a polymer. (Right) Two-cavity cell, for measuring diffusion through a polymer sample. (When appropriate, the liquids in the cell are continuously gently agitated by placing the cell on a rocking platform.) Reproduced with permission from *Food Manufacture*, May 1979.

simulants. For this reason, workers at PIRA used a model compound containing magnesium, since this element is readily determined by atomic absorption spectrophotometry. Other organic compounds are not so readily estimated even with modern analytical techniques, such as gas–liquid chromatography, gas chromatography/mass spectrometry or high-pressure liquid chromatography. Hence, many experiments, particularly by Figge[9] and co-workers in Germany, have employed radiochemically labelled compounds. Conventional chemical techniques for the determination of specific migrants in aqueous, alcoholic and fatty simulants have been extensively reviewed by Crompton.[24] Whilst the analytical stages of the experimental work are considerably helped by the use of radio-tracer methods, other problems are created and must be overcome. First of all, the additive under investigation has to be prepared and labelled radioactively using the standard methods of organic synthesis. In most cases compounds are labelled with either carbon-14 (^{14}C) or tritium (^{3}H). The next stage is to determine the specific activity of the compound and then to incorporate a known weight of it in the plastic under test. Finally, the product must be examined to ensure that the labelled compound is distributed evenly both over the surface of the plastic and throughout its thickness. Then specimens can be tested in special cells filled with a known extractant, or food, under controlled conditions. Measurements of the radioactivity of the extractant are made during the test, or at the end of the test period, and are directly proportional to the mass of labelled compound transferred.

A number of different devices have been described for controlling the contact between plastic and food or simulant. Such migration cells are designed so that the test specimen comes into contact with the extractant on both sides, or one side only. To reduce the analytical problems posed by low limits of detection, the cells are designed to maximise the ratio; surface area of plastic to volume of extractant. This ratio has only a small effect on the migration process except when the concentration of additive is very high. Figure 6 shows both a single cavity cell and a double cavity cell, designed by workers at PIRA.

The measurement of specific migration has contributed greatly to our understanding of the fundamental process of transfer of an additive from a plastic into foods. It has enabled the effects of parameters such as time, temperature, extractant, additive, plastic, thickness, etc., to be evaluated and assessed as to their importance in the design of a test, for the control of global migration and for the protection of the consumer.

REFERENCES

1. Paul, A. A. and Southgate, D. A. T. (Eds.), *McCance and Widdowson's the composition of foods*, 4th edn, *MRC Special Report No. 297*, HMSO, London, 1978.
2. Coppock, J. B. M., Natural toxicants and food additives, *Nutrit. Bull.*, 1979, **5**, 145.
3. Nursten, H. E. and Williams, A. A., Volatile constituents of the blackcurrant, *Ribes nigrum* L., *J. Sci. Food Agric.*, 1969, **20**, 613.
4. Coleman, R. L. and Shaw, P. E., Analysis of Valencia orange essence and aroma oils, *J. Agr. Food Chem.*, 1971, **19**, 520.
5. Chemical Rubber Company, *Handbook of food additives*, Furia, T. E. (Ed.), 2nd edn, 1979.
6. Rossi, L., Interlaboratory study of methods for determining global migration of plastic materials in liquids simulating fatty foodstuffs, *J. Assoc. Offic. Analyt. Chem.*, 1977, **60**, 1282.
7. Figge, K., Private communication.
8. Bishop, J. R., Considerations of how representative are present rather simple extraction tests, *Annali Ist. sup. Sanità*, 1972, **8**, 337.
9. Figge, K., Migration of additives from plastics films into edible oils and fat simulants, *Food Cosmet. Toxicol.*, 1972, **10**, 815.
10. Figge, K. and Koch, J., Effect of some variables on the migration of additives from plastics into edible fats, *Food Cosmet. Toxicol.*, 1973, **11**, 975.
11. Figge, K. and Freytag, W., Vergleich der Additivübertritte aus Kunststoffe in Prüffett HB307, Oliven-und Fischöl, *Dtsch-Lebensmittel Rdsch.*, 1978, **74**, 363.
12. Woggon, H., Uhde, W. J. and Zydek, G., Beitrag zur Prüfung von Bedarfsgegenständen aus Plasten, *Z. Lebensmitt.-u Forsch.*, 1968, **138**, 169.
13. Hofmann, W. and Ostromow, H., *Kautschuk u. Gummi-Kunststoffe*, 1972, **25**, 145.
14. Woggon, H., Uhde, W. J., Romminger, K. and Hoppe, H., *Nahrung*, 1974, **18**, 671.
15. Pallière, M., Contribution à l'étude de la détermination de la migration globale des matérieux destinés a l'emballage des denrées alimentaires ayant un contact gras, *Annali Ist. sup. Sanità*, 1972, **8**, 365.
16. Van Battum, D. and Rijk, M. A. H., The use of Fettsimulans HB307 for the determination of the global migration in fatty foodstuffs, *Annali Ist. sup. Sanità*, 1972, **8**, 421.
17. Koch, J. and Kröhn, R., Neue Verfahren zur Bestimmung der Gesamtmigration aus Kunststoffen in Fette, *Dtsch-Lebensmittel. Rdsch.*, 1975, **71**, 293.
18. Figge, K., Determination of total migration from plastics-packaging materials into edible fats using a ^{14}C-labelled fat simulant, *Food Cosmet. Toxicol.*, 1973, **11**, 963.
19. Rossi, L., Sampaolo, A. and Gramiccioni, L., Méthodes de détermination de la migration globale dans les gras, *Annali Ist. sup. Sanità*, 1972, **8**, 432.
20. Piacentini, R., Outline of a gas chromatographic method for the determination of total migration of additives from rubber goods into fats, *Annali Ist. sup. Sanità*, 1972, **8**, 410.

21. MATURI, V. F., WINKLER, W. O. and YATES, W. E., Determination of and data on total extractables from paper and film by food-simulating solvents, *J. Assoc. Offic. Analyt. Chem.*, 1962, **45,** 70.
22. VAN BATTUM, D., Round-robin analysis on the determination of global migration into fatty foodstuffs, *Report No. 4925,* CIVO-TNO, Utrechtseweg 48, Zeist, Netherlands, 1976.
23. FIGGE, K., CMELKA, D. and KOCH, J., Problems involved in and a comparison of methods for the determination of total migration from packaging materials into fatty foods, *Food Cosmet. Toxicol.*, 1978, **16,** 165.
24. CROMPTON, T. R., *Additive migration from plastics into food*, Pergamon, Oxford, 1979.

Chapter 8

OTHER FOOD CONTACT MATERIALS

INTRODUCTION

Whilst the majority of this review has been concerned with the composition and use of *plastic* containers, utensils and packaging materials in contact with foods, other materials such as metals, ceramics, paper and rubber are also widely used in food contact applications. Some of the problems specific to the use of these other materials in contact with foods will now be discussed.

METALS

Cans (Tin Plate)

Correctly processed canned foods remain microbiologically stable and organoleptically acceptable for long periods. Thus, a can of veal produced in 1824 for Captain Parry's expedition in search of the North West Passage, was opened in 1938 and found to be in perfect condition.[1] Generally, however, foods are canned so that excessive quantities of fresh foods that cannot be eaten immediately following harvest, are preserved for consumption at a later date, when fresh foods are no longer available. Most canned foods, are, therefore, consumed within one to two years of processing. Whilst some canned foods have been found to be fit to eat after many years, changes do occur on storage and there is a progressive loss in subjective values of quality such as colour, flavour, texture, etc. Some nutritional changes may also occur, as well as interactions between the container and the food. Most canned foods contain a high proportion of water in the form of liquids, e.g., fruit juices, beverages or solids packed in a syrup, brine or sauce. In such cases, heat treatment is necessary after canning to destroy spoilage organisms. A few foods, however, are canned in the dry state and require no heat treatment after canning. Canning, in such cases, prevents loss of volatiles from say roasted coffee, or prevents pick-up of moisture by hygroscopic powders (e.g. milk powder). As a further protection against deterioration, dry-pack foods may be hermetically sealed under partial vacuum, or sealed in an atmosphere of an inert gas such as nitrogen. This helps to prevent the development of off-flavours through oxidation of fatty acids (rancidity).

The quality deterioration of foods in cans obviously depends on processing and storage conditions. Colour change is perhaps the most apparent visual indication of deterioration for fruit and vegetables, as well as gas production leading to swollen cans. Such cans may be 'springers' in which the ends can be pressed in and out against the internal gas pressure, or 'hard swells' with permanently deformed ends. Hard blown cans require bacteriological examination to determine the nature of the spoilage. This may have been caused by insufficient thermal treatment after canning or it may be the result of spoilage during cooling by ingress of organisms present in the cooling water through defective seams. Examination of canned foods normally includes physical testing of the vacuum and headspace as well as inspection of the condition of the can contents, the can itself (especially the can seams), together with chemical and bacteriological analysis of the contents where necessary.

Few changes in the nutritional qualities of canned foods will occur on storage since any heat labile nutrients will have been destroyed during thermal processing. In the case of fruits, the low pH value resulting from acids, naturally occurring in the fruit, inhibits the growth of bacteria and only yeast and moulds need to be destroyed. This requires the minimum of heat treatment. Meat products, on the other hand, may contain organisms that produce spores which have a much greater resistance to heat treatment. Heating under pressure (i.e. at higher temperatures) results in greater deterioration of the food, particularly in large cans which have to be heated for longer periods to ensure adequate heat transfer to the centre of the can. Hence, the can serves as a mini pressure cooker, primarily to kill-off all harmful micro-organisms and hence, prevent the putrefaction which would otherwise take place if the food was 'untreated'. Table 1 shows the pH values of some fruits and vegetables after canning.

Food–Can Interactions

Most cans are made from sheet steel coated with a very thin layer of tin (less than 0·5 mm). The can body is formed by bending the sheet into a cylinder

TABLE 1

pH VALUES OF FRUITS AND VEGETABLES AFTER CANNING

Comment	pH	
No bacterial growth	2·5	
	—	
Processing	—	
at	—	
boiling	—	—plums
point	3·0	————gooseberries
	—	—prunes
(Yeasts	—	————rhubarb, apricots
and	—	————————apples, blackberries
moulds	—	—sour cherries, strawberries
grow)	3·5	————————peaches
	—	—raspberries
	—	
————————	—	—sweet cherries
	—	—————pears
Limited	4·0	
bacterial	—	
growth	—	—tomatoes
	—	
	—	—tomato soup
	4·5	
	—	
————————	—	
	—	
	—	
Bacterial	5·0	—carrots
growth	—	
and	—	—turnips, cabbage
spoilage	—	————parsnips, beets, string beans
	—	—beans in tomato sauce
	5·5	————————spinach
	—	—asparagus, cauliflower
Processing	—	————haricot beans in brine
must be	—	—broad beans
under	—	————Lima beans
pressure	6·0	
at	—	
elevated	—	—peas
temperatures	—	———corn
	—	
	6·5	

and closing the side seam with solder. Solders may contain up to 95 per cent of lead alloyed with tin, although pure tin solders are used in some cases, e.g. with baby foods. The ends of the can are attached by a double seam machine without the use of solder. Excess solder inside the can is removed by a rotating mop, which may produce 'splashes' in nearby cans. This may account for the considerable variation in lead content found in individual cans of a given product, from the same batch, stored under apparently identical conditions. The tin coating is applied to protect the steel from corrosion by the acidic aqueous contents. It functions by preferential dissolution, i.e. as a sacrificial anode. Thus, most canned foods contain considerable quantities of tin. There is no legal limit on the amount of tin in foods in the UK, but a value of up to 250 ppm is normally considered satisfactory. Additionally, some cans are lacquered internally, as well as externally, using natural or synthetic resins, to reduce attack from corrosive acids present in some fruits, e.g. rhubarb, plums, loganberries and gooseberries, which have natural pH values below 3·0. Further need for lacquering arises from the discoloration of such products that occurs as a result of reaction between the natural colourants (anthocyanins) in red or purple fruits and trace amounts of tin or iron dissolved during corrosive attack. Some vegetables, e.g. peas become grey on storage and synthetic colouring matter is added to produce a commodity acceptable to the consumer. Foods that contain sulphur produce staining and blackening of the can interior, although this effect can also be minimised with a suitable lacquer. Despite lacquering, flaws in the coating may expose greater or lesser amounts of the soldered seam to corrosive attack, and the lacquer may also fix solder splashes that would otherwise be removed during cleaning operations. All of these factors contribute to inter-can variability as described by Thomas *et al.*[2] However, the protection afforded by lacquering is not always complete. Pinhole leaks in the lacquer lead to pitting corrosion, and whilst lacquering reduces the attack on tin, the amount of lead dissolved can increase as tin will then no longer act as a sacrificial anode. Corrosion of iron can affect the quality of the product and through the production of hydrogen gas can cause swelling and ultimately perforation of the can. This increases the amount of lead and iron dissolved, whilst in non-lacquered cans iron is usually only attacked after extensive detinning has occurred. The rate of detinning is also influenced by the nitrate content of the food and associated liquor.[3]

Thomas *et al.*[2] analysed a number of canned fruit and vegetable products and quoted an overall mean for lead in lacquered cans (80 samples) of 1·45 ppm compared with 0·46 ppm (88 samples) for plain cans. Both of these figures are significantly higher than the natural levels found in fresh foods. Results obtained at the Laboratory of the Government Chemist confirmed these findings[4] and, in addition, illustrated a disproportionate concentration of metals in the solid contents of the can in comparison with the liquid (brine or syrup) as shown in Tables 2 and 3.[5] This phenomenon was first reported for tin by Adam and Horner in 1937,[6] but it appears that lead and iron are similarly concentrated in the solid food, possibly by the formation of complexes between the metal and naturally occurring

TABLE 2

DISTRIBUTION OF TRACE METALS IN THE FRUIT AND SYRUP OF CANNED FRUIT
(From Annual Report of the Government Chemist, 1974; HMSO, London, 1975)

Fruit	Fraction mass (g)[a]	Can[b]	Metal content (mg/kg)		
			Lead	Tin	Iron
Mandarin oranges	S 120	P	0·16	130	6·8
	F 130		0·3	180	6·9
Grapefruit	S 240	P	0·18	75	5·3
	F 260		1·00	130	6·9
Plums	S 250	P	0·15	55	5·3
	F 250		0·45	110	4·2
Pears	S 220	P	0·15	130	5·0
	F 250		0·75	120	5·1
Peaches	S 240	P	0·13	35	1·8
	F 260		1·15	80	6·1
Rhubarb	S 220	La	0·16	20	3·0
	F 260		0·30	30	24
Blackcurrants	S 170	La	1	45	1 300
	F 145		10	160	2 600
Raspberries	S 300	La	0·07	10	2·0
	F 260		0·75	20	8·4
Gooseberries	S 265	P	0·08	95	3·0
	F 325		0·30	260	4·3
Strawberries	S 290	La	0·05	5	4·2
	F 210		0·60	20	10·0
Black cherries	S 195	La	0·04	5	3·2
	F 244		0·30	20	11
Apricots	S 190	P	0·09	100	3·1
	F 245		0·30	80	10
Pineapple	S 140	P	0·12	55	31
	F 160		0·35	105	3·0

[a] S = syrup; F = fruit.
[b] P = plain; La = lacquered.

TABLE 3

DISTRIBUTION OF TRACE METALS IN CANNED VEGETABLES

(From Annual Report of the Government Chemist, 1974; HMSO, London, 1975)

Vegetable	Fraction mass (g)[a]	Can[b]	Metal content (mg/kg)		
			Lead	Tin	Iron
Green beans	L 142	La	0·10	5	2·8
	V 170		0·70	10	4·8
Broad beans	L 94	P	0·06	5	6·8
	V 218		0·02	10	18
Haricot beans	L 197	La	0·02	5	9·8
	V 230		0·15	10	26
Processed peas	L 122	La	0·22	10	9·9
	V 190		0·30	20	17
Petit pois	L 67	La	0·04	10	10
	V 156		0·55	20	12
Mushrooms	L 90	P	0·01	15	5·1
	V 122		0·04	55	16
Celery hearts	L 60	La	0·13	10	4·0
	V 280		1·50	20	3·4
Sweetcorn	L 157	La	0·04	10	1·0
	V 265		0·30	20	6·4
Asparagus	L 113	La	0·21	25	57
	V 192		0·05	54	59

[a] L = liquor; V = solid.
[b] P = plain; La = lacquered.

chelating compounds in the foodstuff. The phenomenon is of little consequence in the case of canned fruits, since both fruit and syrup are normally eaten. With vegetables, on the other hand, many consumers reject the brine and heat the product in water before draining and serving. The uptake of lead is known to increase once the can has been opened, presumably as the result of the ingress of oxygen. This effect is more marked at lower pH values and for unlacquered cans. A greater increase in the lead content of fruit juices was observed on storage at 4 °C than at room temperature. It has been estimated[7] from overall mean values, for lead in canned fruit and vegetables, that the mean weekly intake from these sources is about 200 μg. Canned meat and fish contribute additional quantities of lead to the national diet. The problem is discussed further below.

Some new can-making techniques have been described and are still under development. Two-piece seamless cans produced from a flat circular disc by drawing and wall-ironing are available. Many cans used for beverages are now made from aluminium. The importance of canned foods stems largely from their convenience to the consumer. Some 6000 million cans are opened each year in Britain and 30 per cent of all fruits and vegetables grown in the United States are processed and marketed in three-piece cans. However, toxicological considerations are likely to have only a marginal effect on the future of the canned foods market.

Stainless Steels

Materials coming into contact with food must be clean and sterile. The surfaces of food processing equipment must be readily cleansed from batch to batch. Many cleansing agents are aggressive, so that materials possessing good corrosion resistance are required. Stainless steels, as the name itself implies, possess many desirable characteristics and can resist attack by concentrated detergents, hypochlorite sterilising solutions and acids or alkalis. In the food processing industry, stainless steel is used in the construction of storage tanks, transportation equipment, machinery, processing vessels, beer barrels, fermentation tanks as well as kitchen equipment, moulds and other utensils. The material possesses high strength and rigidity as well as resistance to pitting. The surface is readily polished and food residues and corrosion products such as tartar in wines, water hardness salts or calcium phosphates and caseinates from milk products, are readily removed. Wines are particularly sensitive to spoilage by small amounts of metallic compounds of iron or copper and stainless steel also exhibits good resistance to corrosion by sulphur dioxide added as a preservative. Davis[8] has investigated the bacterial cleansing of various materials, used in food contact applications, after the surface had been soiled by aged raw milk. The results are shown in Table 4.

Chemically, stainless steels differ from plain carbon and lower alloy steels by the addition of chromium and nickel. Other elements, e.g., molybdenum, aluminium, titanium and silicon are often included for particular uses. A typical composition would contain 18 per cent of chromium, 8 per cent of nickel and 0·1–0·2 per cent of carbon. Two to four per cent of molybdenum improves resistance to pitting but has little effect on chloride stress-corrosion cracking.

Foods cooked in stainless steel utensils have been found to contain only minute amounts of trace elements such as chromium, nickel or molybdenum and no deterioration in the organoleptic qualities can be detected. Stoewsand *et al.*[9] reported a study of a variety of acidic foods which had been in contact with stainless steel surfaces during harvesting, processing and/or preparation for market. They analysed the foods for the

TABLE 4
NUMBER OF BACTERIA LEFT AFTER CLEANING SURFACES SOILED BY AGED RAW MILK

Surface	Bacterial counts[a]
Glass	2·8
Stainless steel	14·2
Vitreous enamel	14·5
Aluminium	25·7
Polystyrene	57·6
Stoved enamel	102
Teak	251
Polyethylene	316

[a] Logarithmic mean bacterial contact counts per square inch remaining after swabbing with 0·25 per cent Na_2CO_3 solution followed by rinsing.

elements chromium and nickel as indicators of the possible release of metallic elements from stainless steel. The foods examined included red cabbage, sauerkraut, honey, vinegar, cheese whey and wine. The pH values varied from 2·8 for vinegar to 4·8 for cheese whey. Only red cabbage contained any appreciable quantity of chromium or nickel, and this may have been caused by the addition of citric acid (a strong chelating agent) during manufacture. Few details of contact conditions between the foods and the stainless steel were available and aliquots of each food were not tested for natural mineral content prior to contact with the stainless steel. However, the levels of chromium and nickel found were not thought to constitute any hazard to consumers, particularly as small amounts of Cr^{III} and Ni are now thought to be essential constituents of the diet.

Aluminium

The inherent advantages of aluminium as a food packaging material are as follows:

(1) Light in weight;
(2) Good resistance to corrosion;
(3) Non-toxic;
(4) Does not discolour food or impart taint;
(5) Easy to work;
(6) Attractive finish, readily decorated and good feel;

(7) Excellent thermal properties;
(8) Non-permeability to moisture and gases;
(9) Readily recycled.

Aluminium cans were first used in Scandinavia and Switzerland for the packaging of fish, milk, meat and vegetables. In the early 1930s, aluminium foil began to displace other foils based on tin and lead. Nowadays, aluminium is widely used both in cans, as foil wraps for dairy products (butter, cheese and milk bottle tops), cigarettes, chocolate, as a component of laminates, as tubes and as formed containers for cooking. It is also used in food processing plant and machinery. The purity of food grades of aluminium is over 99 per cent. (Some manganese may be present in rigid containers.) Aluminium possesses a tensile strength of 9 to 12 tons per square inch, depending on gauge, and has a specific gravity of 2·7; approximately one-third that of steel. This strength per weight property coupled with good resistance to corrosion will ensure a continued use for aluminium in food packaging, particularly with the growth of frozen and convenience foods. Aluminium possesses excellent thermal properties as the bright surface reflects around 90 per cent of heat radiation falling on it. This helps to suppress any undesirable rise in temperature of the contents during short-term fluctuations. Equally, it permits a rapid extraction of heat from the contents during freezing, as well as quick and even cooking on reheating. Some caution is advisable, however, in the use of aluminium packages in microwave cookers.

In general, the metal displays excellent resistance to a wide range of corrosive agents. Acidity is the main controlling factor. The pH of the contents should be above 5·2 unless the surface is protected. Aluminium cans are not seriously attacked by soda water, fruit juices, or beer, although some problems have been noticed with cider, some wines, spirits and liqueurs. Protective coatings reduce the attack in most cases. Oxalates (present in rhubarb and spinach) accelerate corrosion much more than hydroxy acids such as tartaric, malic or citric. Chloride possesses a strong corrosive action leading to pitting, as does the presence of copper. Sugar appears to reduce corrosive action in products such as jam. Aluminium does not produce dark sulphur stains with vegetables such as peas and turnips. The cans are subject to very little external corrosion, particularly if sodium silicate is added to the process water. They are more readily opened than tinplate containers but will not withstand hydrogen swells to the same degree. A certain amount of aluminium may dissolve and be eaten with the food. Lopez and Jimenez reported[10] that sardines packed in aluminium containers could contain from 85 to 103 ppm of aluminium. Such concentrations are not thought to be harmful as the metal is not absorbed in the alimentary tract. Moreover, small amounts of aluminium are found naturally in foods such as green beans, strawberries, radishes, tea and cabbage leaves. Such trace levels of aluminium have no noticeable effect on the colour, vitamin content or flavour of the product. For some highly pigmented fruit, a slight bleaching effect has been observed.

The use of aluminium foil for the protection of dairy products stems from its properties as a barrier to air, vapours, moisture, bacteria and light.

The thicknesses of metal used vary from 0·009 mm for fats to 0·018 mm for household foil and up to 0·03–0·12 mm for pleated foodstuff containers and pie dishes. The very thin aluminium foils have little strength and, hence, are usually reinforced by bonding to film such as cellophane, cellulose acetate, polyethylene, vinyl chloride, vinylidene chloride, polyesters or paper. Composite products such as the retort pouch usually contain a layer of aluminium for its barrier properties. Aluminium is readily formed into laminates with a wide range of other substances, thus enabling the packaging manufacturer to combine the virtues of several different materials in a single product.

No one packaging material will dominate the field of packaging in future years. Three major considerations influence the selection of the best material namely, technical suitability, appearance and cost. The latter parameter is likely to prove decisive over the next few years.

Pewter

The British pewter industry dates back several hundred years and even today is largely a craft industry. The principal articles of pewterware likely to come into contact with food are beer tankards, wine goblets and, on a smaller scale, flasks, jugs, tea and coffee sets, plates or trays. Originally, pewterware was fashioned from an alloy of tin and lead. Compositions varied from 1 part of lead to between three or up to nine parts of tin. Modern pewterware is an alloy of tin, together with small amounts of antimony and copper. Traditional pewterware, containing large amounts of lead, possessed a dull, almost black, finish whilst modern low-level lead alloys are much brighter in appearance. However, some reproduction antique pewterware is being produced using modern materials that have been darkened by chemical treatment. Hence, the transition towards low-lead containing materials together with the limited opportunities for food contact, both act as a further protection against a hazard which has been very small for many years.

Standards governing the chemical composition of pewter and solders for the manufacture and joining of pewterware can be found in BS 5140:1974.

CERAMICS

The word 'ceramic' is used to describe any product obtained by the action of high temperatures on inorganic materials, only metals and their alloys are specifically excluded from the definition. Modern-day ceramics encompass a great variety of materials ranging from dense, polycrystalline, glass-free refractory materials, through crystalline aggregates bonded in a glassy medium, to wholly vitreous materials such as glasses. Prior to heating the materials are moulded into the desired shape. On heating, hardening occurs and further shaping by mechanical means is not then readily achieved, in contrast to other materials such as metal, plastics or wood. Ceramics are characterised by their great resistance to chemical attack, even at very high temperatures, and by their brittleness under load or stress. Pottery is perhaps the most important example of a ceramic material used in contact with food. Ceramic glazes, vitreous enamels and glasses are discussed in a separate section, *vide infra*.

The fundamental physical and chemical properties of ceramics have been described by Salmang.[11] Silica is the major constituent of all ceramics. Clays, another major raw material, consist chemically of three components Al_2O_3, SiO_2 and H_2O, in varying proportions. Feldspar is used as a flux and consists of sodium, potassium or calcium aluminosilicates. There are two main types of clays, viz. kaolins (including china clay) and other clays. Kaolins are white and remain so on heating in contrast to other clays, which are reddish-brown in colour as a result of the presence of soluble compounds of iron. The main properties of clays are their plasticity, or workability in the natural state, and their refractory properties on heating. Whilst the principal structural unit is undoubtedly the silicon tetraoxide (SiO_4) tetrahedron (Fig. 1) other tetrahedral, cubic, hexagonal and octahedral arrangements may also be present, especially where compounds such as silicides, borides, nitrides and carbides are involved in addition to the silicates. The silicon–oxygen tetrahedra $[SiO_4]^{4-}$ are linked together through the adsorption of water molecules on the negatively charged oxygen atoms, or directly through the free oxygen valencies, to other $[SiO_4]^{4-}$ tetrahedra, or to two $[AlO_6]$ octahedra, etc. Thus, a network principally consisting of silicon–oxygen tetrahedra is obtained, but modified as a result of the different size, ionic radii and co-ordination

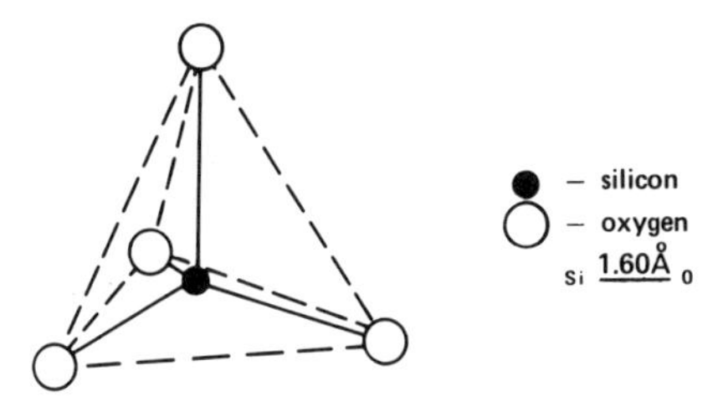

FIG. 1. The silicon–oxygen tetrahedron.

patterns of other elements that may be present. Adjacent layers are stabilised additionally by the action of weak van der Waals' forces. In addition to chemical considerations, physical properties, especially particle size, can have an important influence on the properties of the product. The particle size can vary from above 0·2 μm down to ultra-microscopic or colloidal dimensions.

Firing

On drying, clay shrinks just as it swells during moistening. Control of the shrinkage process is most important in the ceramic industry. Water is thought to occur in clay in three different forms:

(a) Water in between the particles;
(b) Water surrounding individual particles;
(c) Water adsorbed onto the external surface of particles.

During drying water is removed in distinct stages corresponding to the mode of combination within the clay. At higher temperatures, the clay substance breaks down into free oxides such as Al_2O_3 and SiO_2. An exothermic reaction is observed in the range 900–1050 °C, possibly the result of a conversion of alumina from the γ- to the α-form, or to the reaction between γ-alumina and silica. Hence, during the firing process, the last traces of mechanically held water, not removed during drying, are expelled. Up to 1000 °C, expansion and contraction processes occur with the expulsion of chemically bound and structural water, leading to an increase in porosity. Other minor changes include the loss of carbon dioxide from carbonates and the oxidation of sulphides. Above 1000 °C, other complex changes occur, producing a reduction in porosity, shrinkage and the development of increased strength. This process is often referred to as sintering, by which a powder mixture is heated near to the melting point in order that the particles will coalesce and bind together without liquefaction. In vitrification, an appreciable quantity of glass phase is obtained through eutectic melting and the separation of a crystalline phase. During the production of glass, complete melting occurs, producing a liquid of high viscosity which on cooling does not crystallise since the high viscosity prevents the necessary atomic rearrangements from taking place.

Glazes

Ceramics are frequently glazed to make the surface smooth and impermeable. Desirable properties of a glaze include:

(1) Insolubility in water, acids and alkalis;
(2) Resistance to scratching, crazing, peeling, etc.;
(3) Fusible at suitable temperatures, impervious and decorative as required.

In chemical composition, glazes are very similar to glasses. Modern glazes are complex mixtures of borates and silicates. The constituents of the mix are melted and then cooled in such a way as to prevent crystallisation. Crystalline particles scatter light, reducing translucency. As with all silicates, the basic structural unit is the silicon–oxygen tetrahedron (Fig. 1).

TABLE 5

IONIC RADII AND POTENTIAL OF SELECTED CATIONS

Cation	Valency	Ionic radius	Ionic potential
Li	1	0·60	1·7
Na	1	0·95	1·0
K	1	1·33	0·75
Be	2	0·31	6·5
Mg	2	0·65	3·1
Ca	2	0·99	2·0
Sr	2	1·13	1·8
Ba	2	1·35	1·5
Zn	2	0·74	2·8
B	3	0·20	15·0
Al	3	0·50	6·0
Si	4	0·41	9·8
Ti	4	0·68	5·9
Zr	4	0·80	5·0
Sn	4	0·71	5·6
Pb	4	0·84	4·8
P	5	0·34	14·7
As	5	0·47	10·6

However, in glassy materials the units are linked together to form a three-dimensional random network. The interstices of the network can be filled with stabilising elements which modify the physical properties of the glass. The high viscosity of the melt prevents atoms or ions ordering themselves in rows, layer upon layer over long distances during the cooling process. There may be some short range regularity so that glass is, in effect, a supercooled liquid. In addition to silicon, other glass-forming elements, e.g. boron, phosphorus, arsenic and germanium are known. These can also form tetrahedral arrangements with oxygen (triangular in the case of boron). Such elements have an ionic potential (valency ÷ radius) of at least seven (see Table 5) indicating that strong ionic bonds are formed with oxygen. Other elements, e.g. sodium, potassium, magnesium, lead, aluminium, zinc and lithium occupy spaces in the network determined by their size and ionic radius (or sphere of action) and are known as stabilising elements. Such elements are relatively large by comparison with glass-forming elements. Lead is of particular importance in both glasses and glazes since its presence confers low melting temperatures, low viscosity and a high refractive index to the glaze leading to a brilliant and smooth surface finish. It is this use of lead that gives rise to the major concern over the use of ceramics as food utensils.

Vitreous Enamel

This is a finish applied to metal, usually cast iron, in the form of a powdered glass using either a dry or wet process. The latter process essentially involves coating the metal by dipping; but in the dry process the metal is first heated above the melting point of the powder. The enamel melts on contact and adheres to it, to form a continuous coating by fusion.

HOUSEHOLD UTENSILS AND ARTICLES

Contamination arising from the use of lead or cadmium in decorative glazes applied to domestic articles used in contact with food has been the subject of legislation in some countries for many years. In the various standards proposed, a distinction is made between different types of articles. The following definitions are taken from The Vitreous Enamel-Ware (Safety) Regulations, 1976:

> '*Cooking ware* means an article any part of the inside of which is coated with vitreous enamel and which is designed to contain food and to be heated in contact with such food in the course of its preparation, and

includes a casserole, saucepan, frying pan, roaster, percolator, steamer and a kettle.

Flatware means an article any part of the upper surface of which is coated with vitreous enamel and which is designed to be used in connection with the serving or consumption of food, being an article of which the ratio of the height to the diameter is less than one-half when calculated as indicated below, and includes any soup, dessert or other plate and a saucer. The height of any article is defined as the vertical distance from the lowest point on the internal hollow surface to the highest plane to which the article can be filled with water without overflowing, when standing on a level surface. The diameter is the internal diameter of the largest horizontal cross-section when such cross-section is circular. When the largest horizontal cross-section is not circular the diameter is the diameter of the circle which is equal in area to the largest horizontal cross-section.

Hollow ware means an article any part of the inside of which is coated with vitreous enamel and which is designed to be used in connection with the storage or consumption of food, being an article of which the ratio of the height to the diameter is not less than one-half when calculated as indicated above, and includes a cup, mug, jug, jar, tea pot or coffee pot.

Kitchen utensil means a utensil any part of which is coated with vitreous enamel and which is designed to be used in contact with hot food in the course of the preparation of the food, and includes a fish slice, ladle and a funnel.'

Metal Release Tests

The Vitreous Enamel-Ware (Safety) Regulations, 1976 SI 1976 No. 454 were made under the Consumer Protection Act of 1961 rather than the Food and Drugs Act, 1955. The Regulations in turn make reference to British Standards No. BS 5180: Metal 1974, Specification for Permissible Limits of Metal Release from Vitreous Enamel-Ware. Part I concerns tableware and Part 2 relates to cooking ware and kitchen utensils. The standards of release for lead and cadmium are shown in Table 6. The extraction test involves a contact period of 24 h at room temperature (19–21 °C) of the ceramic article with 4 per cent acetic acid in the absence of light. Metals released into the extracting solution are then estimated by using atomic absorption spectrophotometry. In the case of cooking ware, the article is left in contact with extractant for 2 h at boiling point, followed by cooling for 22 h to room temperature before analysis.

TABLE 6

PERMISSIBLE LIMITS OF METAL RELEASE FROM VITREOUS ENAMEL-WARE

	Lead (mg/litre)	Cadmium (mg/litre)
Tableware: hollow ware	2·0	0·2
flatware	7·0	0·7
Kitchen utensils	50 mg/m^2	5 mg/m^2

(From BS 5180 Parts 1 and 2: 1974)

Similar proposals are currently under discussion in the EEC although there may be some changes in the limits of extractable metals finally approved. There are also differences in the definition of ceramic articles and a separate category is proposed for those utensils designed specifically for use by very young children, since they may be hypersensitive and, hence, require greater protection. Packaging and storage containers are also separately defined since prolonged contact between the food and the article is likely.

Many standards and testing procedures for acid-extractable lead and cadmium have been devised in different countries throughout the world. For example, citric acid has been used in place of acetic acid as an extraction agent. The strength of acetic acid used can vary from 3 to 4 per cent. Differences in temperature and time of test, the method of exposure and the extract in the technique used to determine lead or cadmium have all been prescribed. Nevertheless, such minor details give rise to relatively small changes in the amount of metallic compound extracted and can be predicted within experimental limits. Of greater importance is the effect of light as a parameter in the test. Halpin and Carroll[12] showed that in the absence of light a particular batch of plates gave values of $0{\cdot}17 \pm 0{\cdot}06$ ppm for extractable cadmium. However, when the test was carried out in daylight, or in artificial light, a value of around 2·5 ppm was obtained, depending on the duration, wavelength and intensity of the lighting conditions used. The extraction of lead was constant at $1{\cdot}2 \pm 0{\cdot}4$ ppm. They attributed the discrepancy to the photo-oxidation of insoluble cadmium sulphide to soluble cadmium sulphate. These initial findings have since been confirmed by other workers, so it is essential that exact lighting conditions are specified in any test to determine the extractability of cadmium, if reproducible results are to be obtained.

Results of Extraction Tests

Whilst the leaching of lead from glazed ceramics used as food and beverage containers has been identified as a potential source of lead ingestion, recent surveys of commercial products have shown that most articles are within the standards set by the Food and Drug Administration (FDA) of the USA or the British Standards Institution, i.e. a maximum of 7 mg of lead per litre of extract after testing under prescribed conditions. Only some home-made domestic products were found to exceed the allowed limit (Henderson *et al.*, 1979).[13] In an earlier survey of imported products, Seth *et al.*[14] found that all 231 samples tested met the FDA limit and were within the range 1·0 to 6·5 mg/litre, with a mean value of 4·2 mg/litre. They found that during repeated extractions on the same sample, the amount of lead leached out decreased with each successive extraction, so that by the fifth extraction the level had fallen to about one-tenth of the initial value. Since the leaching test is designed to represent extreme conditions of domestic use, under normal conditions any lead released into food is likely to occur over a long period of time and to decrease slowly with repeated use of the utensil. The importance of temperature was also demonstrated. Up to 70 °C there was only a gradual increase in the amount of lead leached into the extractant. Between 70 and 90 °C, however, the amount of lead released increased sharply as illustrated in Fig. 2.

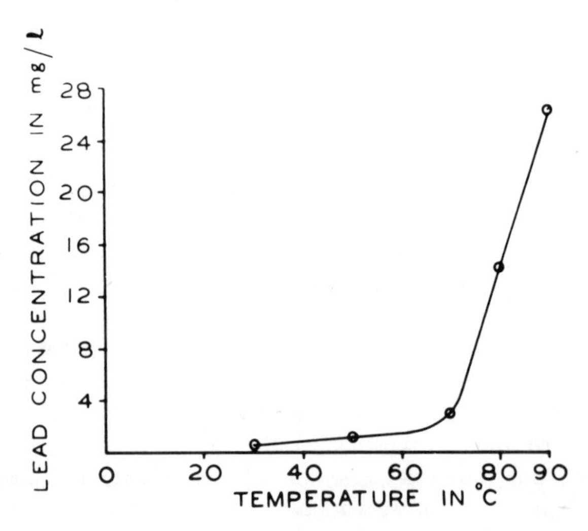

FIG. 2. Effect of temperature on the extraction of lead by 4 per cent acetic acid in 6 h from pottery.

Further protection from lead poisoning is afforded by the Cooking Utensils (Safety) Regulations 1972 (SI No. 1957). These state that:

> 'No kitchen utensil in which to cook food shall be coated, on any surface designed to come in contact with the food, with a tin or other metallic coating which (apart from any local contamination derived from solder forming part of the utensil) contains lead or any compound of lead so that the proportion by weight of lead calculated as the element (Pb) exceeds 20 parts in 10 000 parts of that coating.'

i.e. not greater than 2000 ppm. Painted items found in the household are also a potential source of toxic metals. Yellow pigments and paints frequently contain lead chromate ($PbCrO_4$) or cadmium sulphide. Red pigments may be prepared oxides of iron or may contain a cadmium sulphide/selenide complex. Barium compounds may be present in green-coloured articles. Alternatively, they may consist of mixtures of yellows and blues, e.g., lead chromate and Prussian blue [$KFe^{III}Fe^{II}(CN)_6 . 14H_2O$]. Organic pigments are also used in addition to, or as a replacement for, inorganic compounds. Hence, additional legislation has been enacted in order to protect consumers (especially children) from the use of toxic paints on items such as toys or pencils. Such statutory instruments lay down maximum permitted levels of metals such as arsenic, antimony, barium, cadmium, chromium^{VI}, lead and mercury, that may be present in the coating or paint used to colour the article.

GLASS

Glass packs are still the most popular form of container for foods like milk, jam, carbonated drinks and fruit juices, as well as some lower sales-volume products such as meat and fish spreads or herbs. Over 80 per cent of milk, 90 per cent of jam and 60 per cent of soft drinks are packaged in this way. Beer, wines and spirits comprise another substantial sector of the glass container market. Further applications include cooking ware and tableware. The chief advantages of glass as a packaging material are its chemical inertness, long shelf life (without denting), strength and appearance, including visibility of contents and surface sparkle. Glass is also easily cleansed and is impervious to odours and flavours. It is, however, readily broken with a low resistance to shock treatment and it will not always withstand sudden changes in temperature. The surface of glass containers is subject to imperfections, scuffs and scratches during mechanical handling, which reduces its overall theoretical strength, although treatments with compounds of tin and selected lubricants can reduce surface damage. However, glass containers are reusable and can be recycled when scrapped. In 1979 in the UK, total sales of glass containers were around 7000 million.

At room temperature, glass is a hard, rigid and brittle solid. It is amorphous and usually clear. As the temperature is raised by several hundred degrees centigrade, glass gradually softens and finally becomes a viscous liquid. Commercially, glass is produced by fusing together sand, soda ash and limestone. A wide range of products varying in physical properties can be obtained by incorporation of other ingredients such as oxides of boron, phosphorus, magnesium, lead, zinc and barium. The viscosity, coefficient of expansion, refractive index and electrical properties are strongly dependent on the composition, whilst thermal conductivity and mechanical strength vary only slightly from one glass to another. Waste glass (or cullet) is usually added during manufacture to assist melting. The cooling process must be very carefully controlled to reduce internal stresses within the highly viscous fluid, without at the same time allowing crystals to form thus reducing the clarity. Glasses used for high quality tableware usually contain relatively high concentrations of lead, producing products of high lustre which can be easily worked. Aluminosilicate glasses containing around 20 per cent of Al_2O_3 are more suitable for high temperature applications, such as cooking ware. 'Pyrex' glass is a borosilicate composition which has exceptional resistance to thermal shock. Coloured glasses have only a limited use in the food industry. Amber glass which suppresses transmission of ultra-violet radiation is used for beer bottles, and emerald green glass for some soft drinks and alcoholic beverages. Some typical chemical compositions are shown in Table 7.

Chemically, glass is highly resistant to attack from water, aqueous solutions and organic compounds. Water and acids have very little effect on silica although they may attack some other constituents of the glass. Alkaline solutions are generally much more corrosive to glass, especially when hot. Standard tests have been developed in which glass containers are autoclaved with various test liquids under defined conditions. Evidence of attack is usually derived either by measurements of loss of weight, or alternatively by chemical analysis of the test liquid for components present in the glass. Silica and alkali are the main components leached from the glass. Initially, the rate of solution varies approximately with the square root of the time, suggesting a diffusion mechanism of leaching. As the

TABLE 7

COMPOSITIONS OF SOME COMMERCIAL GLASSES (Values in per cent)

Type	SiO_2	Al_2O_3	Na_2O	K_2O	MgO	CaO	BaO	PbO	B_2O_3	Fe_2O_3	SO_3	F
Colourless bottle, dolomite type	73·3	1·93	14·07	0·46	3·76	5·82	0·34	—	0·04	0·37	0·23	0·06
Colourless bottle, calcite type	72·3	1·86	13·46	0·54	0·17	10·63	0·55	—	—	0·60	0·31	0·25
Amber bottle glass	72·5	1·45	13·71	0·61	3·77	7·34	0·29	—	—	0·087	—	0·26
Emerald green	72·5	1·48	15·16	0·34	3·62	5·99	0·35	0·13[a]	0·17	0·19	0·21	—
Light lead crystal	67·2	—	—	7·1	9·25	0·9	—	1·48	—	—	—	—
Pyrex	81·0	2·2	3·6	0·2	—	—	—	—	0·13	—	—	—

[a] This value was obtained with Cr_2O_3 also present.

calcium oxide content of the glass increases, the weight of material leached decreases.[15] Acidic solutions tend to extract a greater concentration of sodium compounds from the glass compared with water alone, although the total migration into acidic solutions is normally much lower than the total migration into water. The main chemicals extracted into aqueous solutions (i.e. silica and sodium oxide) are unlikely to have any significant effect on the organoleptic properties of foodstuffs. Silica is a natural constituent of many foods and some silicon containing compounds are permitted food additives for use as anti-caking or anti-foaming agents. In the UK, the use of such compounds is controlled by the Miscellaneous Additives in Food Regulations, 1974.[16] Levels of such compounds are determined by manufacturing requirements and are generally very high in comparison to the amount of silica derived by leaching from glass containers. Leaching tests for toxic metals such as lead or cadmium have been published as British Standards on similar lines to those developed for glazed ceramic ware. However, lead and cadmium seldom occur in glasses used in food contact applications apart from the use of lead in high quality crystal ware. The danger of contamination by leaching of these metals from glass into food is, therefore, extremely remote. Where lead is present the high temperature fusion process used during manufacture effectively prevents the release of significant quantities into foodstuffs.

Regulations in force in other countries throughout the world do not differ greatly in principle from UK standards. Test conditions generally employ 4 per cent acetic acid at room temperature and separately at a higher temperature for varying periods of time. The extract is analysed for metals such as lead, cadmium and zinc as well as fluorine. In some regulations, limits for arsenic and antimony are also specified (see Table 2, Chapter 5).

LEAD POLLUTION IN PERSPECTIVE

Problems discussed so far regarding the use of various metals and ceramics as food packaging materials have been concerned mainly with the extraction of a number of different toxic metals such as lead, cadmium, arsenic, mercury, antimony, barium, chromium and nickel from domestic utensils. Of these, the greatest concern exists over lead. It is right, therefore, at this stage to consider, briefly, the wider aspects of lead pollution so that the subject is seen in proper perspective.

Lead can be readily extracted from mineral deposits such as galena (lead sulphide). The metal has a relatively low melting point (327 °C), does not readily corrode and is malleable. Hence, it has found widespread use in a variety of different applications even from ancient times. Nowadays, some 240 000 metric tons are used annually in the UK. The principal uses of lead are in the manufacture of batteries and as an anti-knock agent in petroleum products. The widespread natural occurrence and use of lead has ensured that it is now dispersed throughout all parts of the environment. Thus, man ingests lead from food and drink and inhales it in aerosol and particulate form from the atmosphere. Whilst most public water supplies contain no more than 0·01 ppm of lead, in certain areas of the UK where the water is naturally plumbosolvent and lead piping and storage tanks are used, much higher concentrations are found, particularly after standing overnight. In some cases even the limit of 0·1 ppm recommended by the World Health Organisation may be exceeded.

Atmospheric Lead

Anti-knock additives in petrol contribute around 90 per cent of the lead in the atmosphere. Lead added to petrol as the tetraethyl compound is expelled in the exhaust gases as the oxide or as miscellaneous halogen derivatives. Whilst there has been a phased reduction of permitted lead levels in petroleum products from 0·84 g/litre in 1971 to around half this value today, this has been offset to some extent by a marked increase in traffic density. Hence, lead concentrations in air are highest in urban areas, and busy city streets may contain around 2 to 5 μg of lead per cubic metre, whilst much lower levels would be encountered in the suburbs and rural areas. In urban situations, the size of particulate matter containing lead varies over the range 0·15–0·4 μm. Larger particles will be filtered out in the upper respiratory tract with the result that the fraction of lead inhaled that is retained in the human body is not easy to estimate. Again, the volume of air inhaled by man will vary from person to person, depending on size and working conditions for example, but is likely to lie in the range 6 m^3, for a young child, up to 24 m^3/day for an adult. Bearing in mind the many uncertainties underlying such calculations, it has been estimated that the maximum likely intake of lead from the atmosphere is around 13 μg/day. Most people would be expected to absorb less than 1 μg/day from this volume of air. However, such figures are no more than ‘guesstimates’ and subject to very wide variations. More reliable data is available on intakes from food.

Lead in Food

In 1971, a working party on the monitoring of foodstuffs for heavy metals was established in the UK following reports in the United States of America

TABLE 8

METALS IN TOTAL-DIET SAMPLES[5]

Metal	Content in the diet (mg/kg) Range	Content in the diet (mg/kg) Average	Calculated intake from food (μg/person/week)	Main contributing foods	Country	Comments	WHO recommended intake (μg/person/week)	Reference†
Lead:								
1972		0·13	1 540	Canned foods, vegetables, cereals, meat, fish	UK		3 000	Working Party Report
1975	—	0·09	1 200	Canned foods, vegetables, cereals, meat, fish	UK	—		Working Party Report
1967			100–2 000		USA			Working Party Report
Cadmium (1973)	0·01–0·04		≯250	Shellfish from certain areas	UK		400–500	Working Party Report
			469		Canada			Kirkpatrick and Coffin
		0·02	350–266	Grains, cereals	USA			Duggan and Corneliussen
	0·027–0·064	0·049	644		USA	Children's institution		Murthy *et al.*
Mercury:								
1973	0·003–0·007	0·005	35–70	Fish, meat, canned tuna, crab, lobster, pig offal	UK	80 per cent as methylmercury	300 total, ≯200 as $CH_3 . Hg^+$	
1971	0·001–0·051		91	Fish, meat, poultry	Canada			Kirkpatrick and Coffin
	0·001–0·041			Fish, meat, poultry	USA			Tanner and Forbes

Selenium	0·15–0·30	—	1 379	Cereals, meat, fish, dairy produce, poultry	Canada	—	—	Thompson *et al.*
Zinc	2·67–6·30	4·75	63 500 118 300	Oysters, liver, meats, dairy products	USA Canada	Children Adults	35 000–150 000	Murthy *et al.* Kirkpatrick and Coffin
Antimony	0·209–0·693	0·361	—	—	USA	Children	—	Murthy *et al.*
Chromium	0·175–0·472	0·331	4 420 1 008	—	USA Canada	Children Adults	140–3 500	Murthy *et al.* Kirkpatrick and Coffin
Cobalt	0·252–0·693	0·556	7 170 350	Fish, leaf vegetables	USA Canada	Children Adults	14 (B_{12})	Murthy *et al.* Kirkpatrick and Coffin
Manganese	0·535–1·639	0·823	11 000 23 000	Tea, bread, nuts, green vegetables	USA	Children Adults	14 000–21 000	Murthy *et al.* Kirkpatrick and Coffin
Arsenic as As_2O_3			525–399	Meat, fish, poultry	USA		0·35 mg/kg	Duggan and Corneliussen
	0·025–0·035 mg/kg body mass		0·049–0·42 mg/kg body mass			—		WHO

† See reference 5.

of high levels of mercury in canned tuna fish. The second report of the working party was concerned with a survey of lead in food and was published in 1972. In this report,[17] the average dietary intake of lead from food was calculated to be 0·2 mg/person/day, or 1·4 mg/person/week. In addition the report emphasises that further lead is ingested from beverages. In 1975, a supplementary report was published.[18] This showed that the average intake of lead had fallen to 1 mg/person/week even with an additional 0·2 mg arising from estimations concerning lead in beverages. Since the World Health Organisation recommend a maximum tolerable level for lead of 3 to 4 mg/person/week, it was concluded that even in 1972 'there was no evidence of harm to health from present levels of lead in food comprising the diet of the average consumer'. Nevertheless, further studies were proposed and recommendations were made that the Lead in Food Regulations (1961) should be changed. Most individual foods analysed contained amounts of lead well below the legal limits. However, some canned foods and vegetables grown in areas of high pollution, or where sewage sludge is used for fertilisation, may contain higher levels than usual.

Similarly, total diet studies have been carried out in several countries throughout the world. A comparison of average weekly intakes of lead from food is shown in Table 8. The Table also shows levels of other toxic metals ingested from food, along with WHO recommended intakes.

Hence, ingestion from food and beverages appears to be the major pathway from the ecosystem into man for lead. It would appear that the atmosphere makes only a minor contribution. Ingestion of lead from glazed pottery and cooking utensils will in most cases be negligible, except where articles intended only for decorative use are inadvertently used for storage or for cooking acidic foods at high temperatures. Home-made domestic ceramics may however present a hazard in some countries (Henderson *et al.*[13]).

Since lead is ubiquitous throughout the environment and both ingested and inhaled by man, its presence in the human body is inevitable. The toxic effects of lead are well known. Acute lead poisoning, as a result of accidental ingestion of *large* quantities, produces symptoms of thirst, burning abdominal pain and vomiting. In severe cases, collapse, coma and death may follow. Chronic lead poisoning from the continuous absorption of very small quantities is more difficult to diagnose and detrimental biochemical changes may be produced before any clinical symptoms appear. Investigations are usually based on levels of lead in the blood. Concentrations of lead in blood up to 80 μg/100 ml are thought to be harmless in adults. Lower levels than this may be serious in children and have been linked with brain and behavioural disorders, although currently this is an area of some controversy. Children may ingest excessive quantities of lead from lead-containing paints or as a result of pica (a craving for unsuitable food) by consumption of soil, dust, etc. In recent years, paint manufacturers have reduced the lead content of their products and legislation has been introduced to limit the lead content to 1 per cent. The whole problem of lead and health has recently been reviewed.[19]

CELLULOSICS

Paper and Board

Paper and paper products have found extensive use as wrapping and packaging materials for very many years and are still of immense importance today. Whilst there is intensive competition from a number of plastic materials, new uses e.g. ovenable boards are currently being developed. Paper itself can only be used in contact with dry foods, but over the years improved papers have become available which are resistant to moisture, gases and fat, as well as possessing enhanced physical properties. Materials used for improvement include paraffin, waxes, synthetic resins, bitumen as well as plastic coating products such as PVC, PVA, polyethylenes, silicones, etc. Furthermore, laminated products of paper, plastic and aluminium are also often used. Such products find wide application in the packaging of milk, fruit juices and other similar products. Boards are usually much thicker than paper and possess greater rigidity and strength but lower flexibility. Paper is one of the few packaging materials which, when carelessly discarded as litter, will decompose by natural bacterial action in a reasonable time. Like glass, a certain amount can be conveniently recycled.

Various grades and qualities of paper are available reflecting the number of different starting materials and manufacturing processes used. Cellulose, wood pulp, rags and waste constitute the basic raw materials used either alone or in combination with each other. Wood pulp is frequently treated with calcium bisulphite or with caustic soda and sodium sulphide. Many other chemicals are used to modify the properties of the final product. Paper is a sheet material in which many small discrete fibres (usually cellulose) are bonded together. The large number of hydroxyl groups present in the molecular structure (see below) contrast with the molecular architecture of the plastic materials, based largely on carbon chains, that have been described previously. This difference also accounts for the high moisture uptake (and hence low moisture resistance) of paper.

Cellulose is a polysaccharide which forms the permanent structural component of plant cell walls. It consists of a macromolecule of at least 100 to 1000 anhydro-glucose units as shown below:

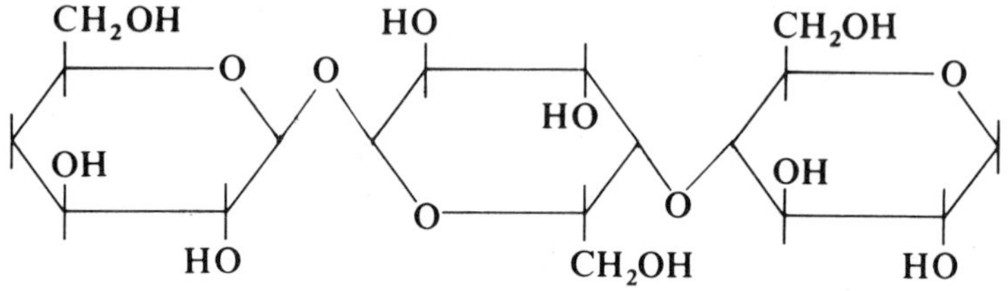

CH_2OH HO CH_2OH
O O O
HO
OH OH
O O
HO CH_2OH HO

However, pure cellulose, e.g. even virgin wood pulp would be too expensive to use on its own and is, therefore, usually used in admixture with other cellulose-containing materials, including waste paper. Hence, there is a possible risk from the use of waste paper as a raw material in food contact applications, that undesirable compounds arising from printing inks, pigments, fillers, preservatives and other additives present in the recycled waste, might thereby come into direct contact with food. Whether any such substance would actually migrate into the foodstuff is still open to question, although with dry foods the effect is more likely to be one of abrasion rather than leaching.

Specialist Paper Products

The name paper is derived from papyrus, a reed found growing near the River Nile and used in ancient times by the Egyptians to produce a writing material. Parchment was the name given originally to dried calf or goatskins that had been treated by rubbing in chalk. Nowadays, substitute materials based on paper are used where specialist properties are required for particular products. Parchment paper (glassine) is highly transparent and translucent and is made from 100 per cent unbleached or bleached treated straw pulp. Greaseproof paper (imitation parchment) is a milled paper, free of wood pulp that has been rendered greaseproof and water-resistant by prolonged beating. Vegetable parchment is a cellulose paper made boilproof and impermeable to fats by treatment with sulphuric acid. Vegetable parchment is often used for wrapping butter, margarine and soft cheeses, whilst glassines are used for confectionery purposes. Waxed papers are used when resistance to moisture is desirable, e.g. for cornflakes and as 'twisting' paper for wrapping sweets. Coated and laminated products are used where one wants to combine the optimum properties of the constituents. For example, very thin aluminium foil is too weak to be used on its own but when bonded to paper gives a combination which is both strong and possesses excellent barrier properties for the wrapping of butter. On the other hand, the use of a coating of a plastic material renders paper usable in heat-sealing applications. An outer layer of paper is often desirable when good surface printability is required.

Regenerated Cellulose

Regenerated cellulose is cellulose precipitated out of solution (including solution as a cellulose derivative), in which any substituent that may have been present has been split off again during the precipitation process so as to regenerate the polyanhydroglucose structure. The cellulose is regenerated in the form of a film which may be uncoated, coated on one or both sides, or as a laminate with other materials such as aluminium, paper or plastics. The cellulose content of regenerated cellulose film is usually greater than 70 per cent depending on the added humectants, slip agents, size and other additives. The problem with this material is, essentially, the very large number of compounds which may be used as additives for which there is incomplete toxicological data.

Cellophane

The name is derived from 'cello' (from cellulose) and 'phane' meaning transparent. It is a thin transparent film consisting of a base sheet of cellulose, regenerated from viscose, containing variable amounts of water and softener. It is usually coated on one or both sides to make it moistureproof and is capable of being sealed with heat or solvent. It is produced mainly from viscose but to a lesser extent by the cuprammonium process. The film possesses fairly good strength and toughness, and is flat and sparkling. Its permeability can be varied as required.

Legislative Aspects

Unlike plastic materials, there are one or two specific references, in the UK legislation, to the use of paper products for wrapping purposes. For example, The Food Hygiene Regulations state:

> 'a person who engages in the handling of food shall not while so engaged use for wrapping or containing any open food any paper or other wrapping material or container which is not clean or which is liable to contaminate the food, and shall not allow any printed

material, other than printed material designed exclusively for wrapping or containing food, to come into contact with any food other than uncooked vegetables or unskinned rabbits or hares or unplucked game or poultry'.

The Preservatives in Food Regulations,[20] 1979 contain a special exemption clause which allows food to contain 'any proportion not exceeding 5 mg/kg (ppm), formaldehyde derived from any wet strength wrapping containing any resin based on formaldehyde or from any plastic food container or utensil manufactured from any resin of which formaldehyde is a condensing component'. This situation arises most frequently in the case of paper products. Vegetable parchment used to wrap dairy products is the subject of a British Standard (1820: 1961). In this standard, limits for impurities such as arsenic, copper, iron, lead, glycerine and other substances are specified.

RUBBER

Rubber is not often used as a food packaging material in its own right. However, it may come into contact with a wide variety of foods through its use as a seal or gasket, as an adhesive and in food processing machinery, e.g. piping, pumps and conveyor belts. A plate heat exchanger used in the milk and beer processing industries might have as much as three and a half miles of rubber seal in contact with the liquid. Other uses for which food grade specifications of rubber are required, include bottle teats, dummies, teething rings, balloons and children's toys.

Natural rubber is obtained by coagulation of the latex obtained from the tree *Hevea brasiliensis*. It occurs as particles dispersed in an aqueous serum, which consists of 95 per cent of hydrocarbons with some proteins, lipids and inorganic salts. Rubber is a polymer based on isoprene (C_5H_8) units as shown below:

$$\left[\begin{array}{cccc} H & & H & H \\ | & & | & | \\ -C- & C= & C- & C \\ | & | & & | \\ H & CH_3 & & H \end{array} \right]_n$$

where *n* is of the order of 20 000. Hence, natural rubber can be regarded as polyisoprene or poly-2-methylbutadiene, in which the monomer units are linked head-to-tail to produce a *cis*-1,4 polymer. The *trans* isomer also exists and is known as gutta-percha.

Rubber possesses quite distinctive properties in comparison with materials (including plastics) discussed so far. It is flexible, elastic, tough, waterproof and impermeable to air. Natural rubber tends to be stiff and hard in cold weather but soft and sticky at high temperatures. This disadvantage has been largely overcome with the development of the vulcanisation process, in which the product is heated with sulphur at temperatures in the range 140–180 °C. This greatly increases both the strength and elasticity of the natural product and there is less variation with change in temperature. During vulcanisation the long flexible polymer chains are chemically crosslinked through sulphur atoms to form a three-dimensional network. This increases the strength and elasticity of the material. Crosslinks are formed roughly every 50 to 100 isoprene units. As the sulphur content increases, and the degree of crosslinking increases, the product becomes harder. Ebonite (or vulcanite) contains up to 45 per cent sulphur and approximately one sulphur atom per 4–8 atoms of carbon on the main polymer chain.

Synthetic Rubbers

The variability of composition, properties and supply of the natural product, particularly during World War II, led to the development of a number of synthetic rubbers. At the same time, attempts were made not merely to duplicate those properties of the natural product that make it so useful, but also to enhance such properties, where possible, giving a range of materials with a wide variety of end uses. Most chemicals used in the production of synthetic rubbers are by-products of the petroleum industry. Ethylene, propylene, butadiene, styrene and acrylonitrile are most frequently used as monomers. Generally, synthetic rubbers possess a less regular chain structure than natural rubber owing to the bulky nature of some of the side chains present. Some of the distinctive properties of these products will be mentioned briefly.

SBR (*styrene and butadiene*)—similar in properties to natural rubber apart from abrasion resistance. Its resilience and elasticity are slightly inferior to natural rubber but it possesses superior heat ageing characteristics.

Butyl—has good resistance to heat, ozone and ageing. It is also resistant to vegetable oils (but not to mineral oils), nitric acid and other cleansing agents.

Neoprene—a polymer of chloroprene:

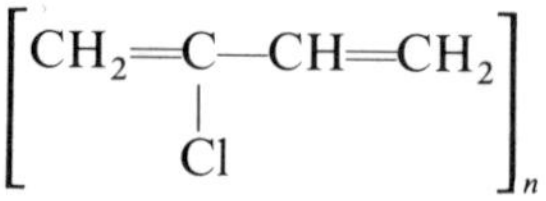

$$\left[\begin{array}{c} CH_2{=}\underset{\displaystyle Cl}{\underset{|}{C}}{-}CH{=}CH_2 \end{array}\right]_n$$

The monomer units are linked 1:4 in the *trans* configuration. On vulcanisation crosslinks are formed with zinc or magnesium oxides reacting with chlorine atoms, as opposed to C—S—S—C bonds in other rubbers. It possesses good oil, solvent and grease resistance. In the presence of steam, some hydrolysis liberating hydrochloric acid can occur. This may cause corrosion of stainless steel plant and machinery.

Nitrile rubber—a copolymer of butadiene and acrylonitrile, with an irregular chain structure. Possesses excellent resistance to solvents and oil.

Silicone rubbers—these are essentially polysiloxanes vulcanised using organic peroxides. The polymer chain consists of silicon–oxygen atoms as shown below:

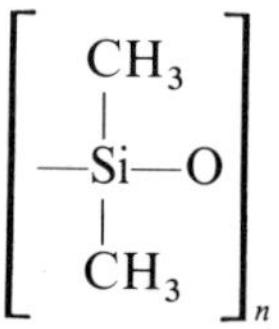

$$\left[\begin{array}{c} CH_3 \\ | \\ {-}Si{-}O \\ | \\ CH_3 \end{array}\right]_n$$

Such polymers exhibit useful properties over the widest possible temperature range (-75 to 260 °C). Fluorinated derivatives possess excellent resistance to oils, and to citrus fruits which are aggressive, as a result of their terpene content, compared with many rubbers and plastic materials.

In many cases the properties of the finished product depend on the particular formulation chosen, the processing, compounding and other treatments used during manufacture. The chemical analysis and characterisation of rubber have been reviewed by Wadelin and Morris.[21]

Legislative Aspects

As discussed earlier for plastics there are no laws in the UK pertaining specifically to the use of rubber products in contact with foods. The Materials and Articles in Contact with Food Regulations only make slightly more explicit what has already been a part of the Food and Drugs Act, 1955. The new regulations apply only to finished products and introduce new labelling requirements. The principal regulations relating specifically to rubber are those issued by the FDA in the United States of America and the BGA recommendations in West Germany. Whilst the latter are not legally binding they are extensively followed in many countries throughout the world.

Substances regulated by the FDA for use in rubber are classified as indirect food additives and are listed in Chapter IB of Title 21 of the Code of Federal Regulations. The following parts are of special importance to the manufacturer:

175.105 Adhesive coatings and components
175.300 Resinous and polymeric coatings
177.1210 Closures with sealing gaskets for food containers
177.2600 Rubber articles intended for repeated use
178.2010 Antioxidants and/or stabilisers for polymers

Generally, accelerators are permitted at levels up to 1·5 per cent and antioxidants up to 5 per cent. Results of water extraction and hexane extraction tests on rubber products were reported by Korte.[22]

The BGA recommendations are contained in Chapter XXI and are divided into different categories depending on the end use of the product as follows:

Category 1

1. Storage containers
2. Container linings
3. Seals with large surface areas exposed to food
4. Sealing rings for cans, glass bottles, etc.

Category 2

1. Food conveying tubes and hoses
2. Stoppers and bungs for bottles and flasks
3. Sealing rings for pressure cookers and coffee machines
4. Seals for milk can lids
5. Ball valves

Category 3

1. Milking installations
2. Milk tubes
3. Seals in dairy installations
4. Membranes and diaphragms
5. Pump stators

6. Roller coverings } For fats and emulsions with fat as a
7. Conveyor belts } continuous phase
8. Rubber aprons and gloves for working with food

Category 4
1. Conveyor belts and roller coverings
2. Suction and pressure hoses for conveying dry food
3. Pipe seals, pump parts, tap washers, all for drinking water

Special Category
1. Children's toys
2. Balloons
3. Bottle teats
4. Dummies
5. Teat caps
6. Teething rings
7. Gum shields

Silicone products are treated separately in Chapter XV of the BGA recommendations. Detailed extraction tests for the different categories are specified. Whilst the German recommendations are somewhat easier to follow, they are more stringent. Inevitably, not many substances find approval under both sets of regulations and there are a number of inconsistencies. The existence of differing legislative controls throughout Europe and the rest of the world presents problems for manufacturers, compounders and end use processors. From this standpoint the trend towards harmonised positive lists initiated by the European Economic Community will no doubt be welcomed by industry providing that such lists and associated legislation is based on commonsense without in any way becoming prejudicial to consumer safety.

REFERENCES

1. HALL, M. N., The shelf life of canned foods, *Nutrit. and Food Sci.*, 1979, July/Aug., 2.
2. THOMAS, B., EDMUNDS, J. W. and CURRY, S. J., Lead content of canned fruit, *J. Sci. Fd. Agric.*, 1975, **26,** 1.
3. FARROW, R. P., LOW, N. T. and KIM, E. S., The nitrate detinning reaction in model systems, *J. Fd. Sci.*, 1970, **35,** 818.
4. HMSO *Report of the Government Chemist, 1974*, HM Stationery Office, London, 1975, p. 40.

5. Crosby, N. T., Determination of metals in foods: A review, *Analyst*, 1977, **102,** 260.
6. Adam, W. B. and Horner, G., The tin content of English canned fruits and vegetables, *J. Soc. Chem. Ind.*, 1937, **56,** 329T.
7. Thomas, B., Roughan, J. A. and Watters, E. D., Lead and cadmium content of some canned fruit and vegetables, *J. Sci. Fd. Agric.*, 1973, **24,** 447.
8. Davis, J. G., *Med. Offr.*, 1963, 8th Nov., 299.
9. Stoewsand, G. F., Stamer, J. R., Kofikowski, F. V., Morse, R. A., Bache, C. A. and Lisk, D. J., Chromium and nickel in acidic foods and by-products contacting stainless steel during processing, *Bull. Environm. Contam. Toxicol.*, 1979, **21,** 600.
10. Lopez, A. and Jimenez, M. A., Canning fruit and vegetable products in aluminium containers, *Food Technol.*, 1969, **23,** 96.
11. Salmang, H., *Ceramics, physical and chemical fundamentals* (Trans. by M. Francis), Butterworths, London, 1961.
12. Halpin, M. K. and Carroll, D. M., Light sensitivity of tests for cadmium on ceramic tableware, *Nature, London*, 1974, **247,** 197.
13. Henderson, R. W., Andrews, D. and Lightsey, G. R., Leaching of lead from ceramics, *Bull. Environm. Contam. Toxicol.*, 1979, **21,** 102.
14. Seth, T. D., Singar, S. and Hasan, M. Z., Studies on lead extraction from glazed pottery under different conditions, *Bull. Environm. Contam. Toxicol.*, 1973, **10,** 51.
15. Rana, M. A. and Douglas, R. W., The reaction between glass and water. I. Experimental methods and observations. II. Discussion of the results, *Phys. Chem. Glasses*, 1961, **2,** pp. 179 and 196.
16. HMSO, *Miscellaneous Additives in Food Regulations 1974*, SI 1974 No. 1121, HMSO, London, 1974.
17. HMSO, *Working Party on the Monitoring of Foodstuffs for Heavy Metals, Second Report*, HMSO, London, 1972.
18. HMSO, *Working Party on the Monitoring of Foodstuffs for Heavy Metals, Survey of Lead in Food: First Supplementary Report*, HMSO, London, 1975.
19. HMSO, *Lead and Health: The Report of a DHSS Working Party on Lead in the Environment*, HMSO, London, 1980.
20. HMSO, *Preservatives in Food Regulations 1979*, SI 1979 No. 752, HMSO, London, 1979.
21. Wadelin, C. W. and Morris, M. C., Rubber, *Analyt. Chem.*, 1979, **51,** 303R.
22. Korte, D. E., Extractables in rubber materials, *J. Assoc. Offic. Analyt. Chem.*, 1967, **50,** 840.

INDEX